A New Geology

Front cover. A possible map of the world 50 million years from now, based on extrapolations of the plate movements discussed in Chapter 14. (After R S Dietz and J C Holden, *Scientific American*, October 1970.)

New School Series

Consulting Editor: **R Stone, MA, A.Inst.P,**
formerly Second Master, Manchester Grammar School

General Editor (Geography and Geology): **M J Bradshaw, MA**
Senior Lecturer in Geology, College of St Mark and St John, Plymouth

A New Geology

M J Bradshaw MA

Senior Lecturer in Geography and Geology,
College of St Mark and St John, Plymouth

HODDER AND STOUGHTON
LONDON SYDNEY AUCKLAND TORONTO

ISBN 0 340 16271 6

First published 1968
Reprinted 1968, 1970, 1972
Second edition 1973
Ninth impression 1986

Printed in Great Britain for
Hodder and Stoughton Educational,
a division of Hodder and Stoughton Ltd,
Mill Road, Dunton Green, Sevenoaks, Kent TN13 2YD
by St Edmundsbury Press Ltd, Bury St Edmunds, Suffolk

Contents

Preface to the First Edition

Geology is a subject which is rapidly increasing in popularity in schools throughout Britain. Perhaps it is surprising that it is still taught in so few, since it deals with a vital part of our environment, which we must understand in order to use properly. Its study provides the opportunity for much practical work on a simple basis and with the minimum of equipment. It is also a discipline leading students to appreciate the value of the scientific method.

The first aim of this book is to cover the GCE Ordinary Level course in schools. It is realised that many students take such a course in the Sixth Form with very little tuition, and the text, end-of-chapter exercises and suggestions for practical work have been written to stimulate the reader to come to grips with the new terminology, and to begin to think in geological terms. The matter treated here is of somewhat wider content than the Ordinary Level syllabuses, and it is hoped that the volume will also be of interest and value to the general reader.

The scope and outline of *A New Geology* is given in Chapter 1. It claims to be 'new' in a number of ways. Firstly it takes the opportunity to introduce the new ideas and discoveries which are yearly resulting from geological research. Secondly, it has a new approach to the basic treatment of the subject at this level, starting with the fundamental dictum: 'The present is the key to the past.' This, and features such as placing a consideration of granites at the end of the chapter on metamorphic rocks, will surely challenge discussion. Thirdly, the whole book places an emphasis of first importance on fieldwork, the practical identification of specimens and map analysis. Fourthly, although much of the material at this level is necessarily culled from textbooks of a more advanced nature and original sources of research, we have tried to include new examples, new photographs and many original illustrations. It is particularly important to think in three-dimensional terms in geology, and block diagrams have been made to feature largely amongst the illustrative matter.

In other words we have tried to produce not another merely factual textbook, but one in which the student is stimulated to think geologically, and to go and find out for himself.

M J BRADSHAW

Preface to the Second Edition

Several factors have prompted the production of a major revision of this book within four years of its first publication.

(1) Geology has become a fast-moving subject after years of results accumulating slowly from painstaking research: new concepts involving major overall aspects of Earth study have come to light and were scarcely mentioned in the First Edition.

(2) Geology has become increasingly important as a school subject: not only have the numbers of GCE Ordinary Level candidates doubled since the First Edition of *A New Geology* was published, but the subject is also being introduced to science courses at all levels.

(3) The author's own view of geology and the ways in which it can be studied have changed with increasing experience and discussion. Geological maps are now introduced as an integral part of the course, and the subject is shown to have economic applications at every turn. The chapter on Fossils has been completely rewritten to provide a closer relationship with the study of living creatures. The section on Earth history is also considerably modified. Perhaps the most important change is the further encouragement given to student participation by the inclusion of a series of investigations at the end of each chapter or section.

(4) The changes involved in metrication also necessitated a substantial revision.

As in the First Edition, *A New Geology* aims to cover the GCE Ordinary Level courses. It has been noted that the matter treated is wider than most individual courses at this level, and the book should be treated partly as a source book and reader, and partly as a basis for the work which can be carried out in the field and laboratory. The First Edition has also been used in CSE, and as a basis for GCE Advanced Level courses in both geology and geography. It is hoped that the Second Edition will continue to meet these varied needs.

M J BRADSHAW

Acknowledgments

The author would like to express his gratitude to the Editor for all the help he has given in so many ways in the writing and revision of this book. Just as he was indebted to Mr I H Forsyth and Dr D I J Mallick of the Institute of Geological Sciences for their reading of the First Edition typescript, so he is equally grateful to his colleague Mr A J Dunk, and to Dr E B Selwood of Exeter University. They have made many valuable suggestions for the improvement of text and diagrams. Several users of the First Edition in schools have also provided constructive comments which have been taken up.

In a book of this nature the author is indebted to many sources, mainly in textbooks of a more advanced nature. These are acknowledged here, and included in the Further Reading List at the end of the book.

The following examination bodies have granted permission for certain of their GCE Ordinary Level questions to be reproduced. They are indicated at the end of each chapter or section by the initial letters in brackets.

The Associated Examining Board (AEB); the University of Cambridge Local Examinations Syndicate (C); the University of London (L); the Joint Matriculation Board of the Northern Universities (N); the Oxford Delegacy of Local Examinations (O); the Southern Universities Joint Board for School Examinations (S); and the Welsh Joint Education Committee (W).

Thanks are also due to the various organisations who have given permission to reproduce photographs: the source of each is acknowledged in the text. Those marked *Crown Copyright* are Geological Survey photographs reproduced by permission of the Controller of HM Stationery Office. Mr John Saunders, of the Geology Department at the University of Exeter, has produced the photographs of rocks, minerals and fossils for the Second Edition.

Ripple-marks. These were once formed on a soft, sandy beach near the Coal Measure forests (ie nearly 300 million years ago). They now form a hard rock surface. (*Crown Copyright*)

Part One

The Earth

1 Geology, the Earth Science

What Is Geology About?

Geology is an exciting subject which concerns many important problems for us as human beings. Man has always depended on the resources of his planet, Earth, to supply his material needs, but he has also asked many questions about it which simply reflect his curiosity of mind. Geologists seek to answer some of these questions:

How and when was the Earth formed?
What is it like inside the Earth?
How has our planet developed to its present state?
How has the scenery we see around us been formed?
What were plants and animals like in the past?
What causes a volcano to erupt?
Where should this dam, or that road, be built?
Where do we start looking for oil, or coal, or uranium . . .?
Where will we find a supply of water?
How long are these resources likely to last?

Geology is thus the science which studies the planet Earth, and it is important for us to try to answer these questions, especially if we are going to be engineers or geographers or biologists, but also if we are to have an intelligent interest in our surroundings and in the future of the human race. The astronauts landing on the Moon require a knowledge of geology in order to bring back an adequate description of what they find and the right samples of Moon rock. The Apollo landings have shown that many of the Moon rocks resemble Earth lavas in their composition: they have also been dated by the methods applied to Earth rocks, and some are older than any rocks at the Earth's surface.

Many geologists would say that the main aim of their studies is to work out the history of the Earth from the time when it first cooled down until man became more or less civilised. Perhaps most of us think that our present landscapes of hills and valleys, lakes and waterfalls, mountains and glaciers, are permanent features, and this is almost true when we consider the rapidly changing events connected with man's progress during the last 6000 years. Geological processes take place so slowly that we can seldom see any great changes in the landscape in a lifetime, but when we start thinking in terms of millions of years we find that Great Britain has been from time to time inundated by the seas, dominated by mountain ranges and deserts, and covered by vast sheets of ice! Geology is indeed 'the story of our changing Earth through the ages', and the knowledge we gain in working out this story is also very useful to the miner and engineer. These practical applications give the study of geology a very realistic significance.

Our Evidence

All scientists base their ideas, hypotheses and theories on a series of facts obtained by observation. The geologist studies the Earth, and most of the facts he observes and records are presented to him at the surface. No one has been to the deep interior of our planet: all our information about that realm has been gathered by indirect observation. The surface of the Earth, however, provides us with a great deal of evidence which we have only just begun to examine. Some parts of the world are known to us in great detail, but most of it has scarcely been visited by geologists. Every now and then we are reminded of this by news of discoveries of valuable deposits of iron or oil in remote parts of the world. Investors can make huge profits when new mineral resources are discovered, as they have done recently in connection with Australian mining concerns: the fluctuating values of the share prices reflect the lack of conclusive geological evidence.

The rocks which are found at the surface provide most of the facts a geologist needs in his study of the past. The minerals which make up the rocks tell us a lot about the conditions in which they were formed,

Suncracks. Those on the left were formed in mudstone nearly 400 million years ago, the cracks having been filled in by the overlying sediments. Those on the right were formed in a period of drought in Kenya during the last year or so. (*Crown Copyright, Aerofilms*)

as do the embedded fossils and the different textures and structures. Some rocks show ripple-mark structures just like those you see on a sandy beach at low tide; others have cracks on the upper surface filled by material from an overlying layer, and these are just like cracks we see in mud dried by the sun; still others contain gas cavities and other features of cooling lava flows, which we find on the slopes of volcanoes throughout the world today. Yet these structures occur in rocks formed millions of years ago and buried under great thicknesses of other rocks. It seems that rain and snow have fallen, the sun has shone and the earth has quaked throughout geological time in much the same

way as they do now. **'The present is the key to the past'** is a basic assumption of modern geology. It can, however, be a rather imprecise catch-phrase. Certainly the distributions of land and sea, the actual types of animals and plants in existence and many other features of the Earth have changed. But it is reasonable to assume that the basic chemical, physical and biological processes have acted in the same way throughout geological time: rates of erosion and deposition have varied, but rivers have taken up and deposited rock debris according to the same principles.

This brings us to the second major group of facts we shall be

This picture shows the results of a lack of adequate knowledge of geological processes: heavy rainstorms have stripped the topsoil and carved gullies in land that should never have been ploughed. This abandoned farm is near Milan, Fort River District, Tennessee. (*Ewing Galloway, New York*)

4

studying, for not only do we have to make a minute investigation into the different types of rocks, but we have to study the geological processes which are in action today. These processes sculpture the landscape, transport rock debris to the sea and deposit it there as the basic ingredient for new rocks, cause upheavals of the land and lead to the eruption of molten rock material from the depths of the Earth.

The Plan Of This Book

The study of these different lines of evidence has determined the outline of this book. We shall begin by considering **the Earth as a planet**. What are its special features? What are the main facts about the surface relief? What are the characteristics of the rocks we find at the Earth's surface, and of the minerals which make up these rocks?

Having taken a broad look at our evidence, and at the planet as a whole, we shall then begin to study each facet in more detail. First comes an examination of the **geological forces** at work on the surface of our planet where the realms of atmosphere, ocean and solid rock meet: rain, rivers, wind, water and ice influence the surface in many ways. This prepares us for a fuller understanding of the **sedimentary rocks**: they are composed of rock fragments and broken-down chemical matter, shell and plant debris. **Fossils**, our source of evidence for the existence of life in the past, occur only in the sedimentary rocks, and they provide the next section of our studies.

There are two other types of rock, igneous and metamorphic, both largely crystalline in nature. **Igneous rocks** have cooled from molten rock material which has migrated from the interior of the Earth. Active volcanoes show us what happens when this material reaches the surface. **Metamorphic rocks** have been subjected to so much heat and/or pressure that the minerals they originally contained have re-crystallised.

All rocks are subjected to forces which deform and fracture them, and as we consider the immense energy involved we shall begin to see a pattern of events at work in the **Earth's history**. In fact we must use the evidence produced by a study of all the various aspects of geology in order to work out what happened in the past. Our final main section will review the principles involved in this investigation, and demonstrate how they can be applied to situations at various stages in the history of the British Isles.

Throughout the course you will be encouraged to apply what you have been learning by reference to your own neighbourhood and places you visit; by interpreting the geological maps, which summarise the geological situation in a pattern of outcrops; by studying the pictures and diagrams in the text; and by looking out for man's use of the resources of his planet.

At the end of the book there are two shorter chapters on some of the techniques you will need to master in order to carry out simple geological work. **Field-work** is the basic method of finding out and collecting information, and can be an extremely effective and satisfying activity. **Map analysis** is important, since you should be able to use one of the geologist's fundamental 'tools', and there is much work which can be carried out only in the **laboratory**. You will probably need to refer to these chapters before you come to the end of the main part of the course. They are designed for this.

As we study each aspect of geology, we shall find that geologists often use the results of other fields of learning, and that it will help us to know some biology, chemistry, physics and geography. Perhaps we shall decide that these various fields, which we often study separately at school, are really just different aspects of our knowledge of the things with which we come into contact and use. Geography together with the physical and life sciences enables us to understand our environment, and to use it to better advantage. Let us hope that our knowledge of geology will add to this understanding, and that it will help us to make more informed decisions and comments in the future. So many tragic mistakes have been made in the past because of ignorance of the facts and ideas you will be learning: it is vital that such mistakes should not be repeated.

2 The Planet Earth

The Earth is one of the smaller planets grouped in the Solar System around the vast nuclear furnace of the Sun. The Solar System, however, is towards the edge of a swirling galaxy (ie a relatively close-spaced group of stars like the Sun) known as the Milky Way, and there are millions of other such galaxies in the universe. Astronomers have been observing the universe for centuries, and, as their instruments have improved, they have found more and more stars at greater and greater distances from the Earth, many of the most distant having been discovered by detecting the radio waves they emit.

Such considerations make us feel insignificant, but present information suggests that our planet is unique in supporting such a complex variety of living forms—at least within the Solar System. It is at least probable that there are other planets which support life amongst so many millions in the universe, but none has been detected with our present instruments. In fact we have not even located another planet as such beyond the confines of the Solar System, since they are too small to be seen. Non-detection, however, is no proof for non-existence. It is most likely that other planets do exist, and that conditions somewhere in the universe are favourable for the development of advanced organisms. We shall probably never meet them!

Before we examine the geological evidence in detail, we must take a look at the broader features of the Earth as we find it today. Amongst the most important factors which determine what we find are the size and composition of the planet, and its distance from the Sun. The first two of these combine to explain the strong gravitational pull exerted by the Earth, which is able to retain a thick layer of atmospheric gases around it. This layer is essential for the maintenance of life. Our

The Earth from space. This photograph was taken by the Apollo 11 crew. Why can you see only part of the Earth's sphere? How are the land, ocean and atmosphere represented? (*USIS*)

distance from the Sun, varying between 145 and 152 million km, means that the atmospheric temperatures at the surface range from −88°C to +58°C, and that a moderate temperature in which life processes can operate (ie 0–40°C) is maintained over most of the surface for most of the time. Few areas are too hot for life, and the polar ice-caps are relatively small. Water exists in liquid form over most of the planet, fulfilling a further basic need of living organisms. The atmosphere also plays a vital part in sifting the Sun's rays, which contain harmful elements capable of burning or irradiating the creatures living at the Earth's surface. In addition gases like oxygen, nitrogen, carbon dioxide and water vapour play important parts in the cycles of organic activity.

The reactions between solar energy and the atmospheric gases cause our weather. Rain, snow, fog, wind and cloud are all due to air and water vapour movements driven by energy from the Sun. These movements help to spread the heat through the atmosphere, and thus to make the planet a place of fewer extremes than it would have been. There are also further reactions, between the atmosphere and the surface rocks, which lead to the destruction, removal and redeposition of these rocks.

The views of the Earth taken from space on the Apollo missions have shown that our planet has a unique appearance. Not only does it have a patchy covering of cloud, but much of the surface is covered by water. The oceans are continuous, isolating the major continental masses, and are particularly dominant in the southern hemisphere. On the other hand this covering of water is a relatively thin film, on average only 3 km deep compared with the equatorial radius of 6378 km. This brings us to the question of the major surface relief, for, of course, the oceans occupy the largest hollows, and the land masses and mountains are merely upstanding masses of rock. The highest mountains in the world are only 20 km above the deepest ocean trenches. (Draw a circle 6 cm in radius to represent the 6000 km Earth radius: your finest pencil line, 0·25 mm thick, is thus equivalent to 25 km, far greater than the Earth's relief, and you could not show any mountains or ocean trenches on such a diagram to scale.)

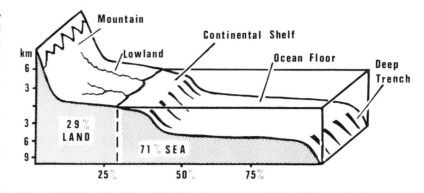

Figure 2.1 The Earth's relief. The importance of the different levels in the relief at the earth's surface is shown. Which two levels occupy the greatest area? The figures at the left of the diagram represent heights in metres above or below sea-level: the highest point is Mount Everest (8848 m) and the deepest the Mariana Trench (−11 003 m). The numbers along the base of the diagram refer to the proportion of the area of the Earth's surface found at the various levels and are expressed as a percentage of the whole.

Figure 2.1 shows that a large proportion of the Earth's surface is just above or just below sea-level. This means that any movement of the Earth's crust which distorts the surface features is liable to alter the extent of water over the continental margins; similarly the freezing or melting of masses of ice affects the sea-level.

Now let us summarise some of the facts and measurements involved.

1 The Earth's Shape

From the measurements given in Figure 2.2 you can see that the Earth is slightly flattened at the Poles. This is related to the fact that it rotates once every 24 hours. Although the Earth appears as a solid ball of

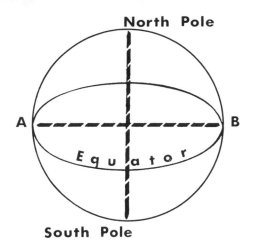

North Pole

A - - - - - - - - - - - - B

E q u a t o r

South Pole

Figure 2.2 The Earth's bulge. The flattening at the Poles is scarcely enough to notice. The Equatorial diameter (AB) is 12 756 km; the line of the earth's axis from North to South Poles is 12 713 km long.

CONTINENT	AREA (km²)	OCEAN	AREA (km²)
EURASIA	54 200 000	PACIFIC	180 500 000
AFRICA	29 800 000	ATLANTIC	92 200 000
NORTH AMERICA	24 200 000	INDIAN	75 000 000
SOUTH AMERICA	18 000 000	ARCTIC	14 000 000
ANTARCTICA	13 100 000		
AUSTRALASIA	9 000 000		
Total land	148 300 000	Total ocean	361 700 000

Figure 2.3

rock, it behaves like a slow-moving fluid over a long period of time, and is slightly deformed for this reason.

2 The Earth's Relief

The most basic distinction amongst surface features is that between land and sea. Compare the areas given in Figure 2.3, and especially the area of the Pacific Ocean with the total land area.

Figure 2.1 shows that the highest mountains and the deepest ocean trenches cover small areas compared with the lower-lying lands and the ocean-floor plains. Some of the highest and deepest points are recorded in the following chart, Figure 2.4.

CONTINENT	MOUNTAIN	ALTITUDE (m)
EURASIA	Everest	8 848
SOUTH AMERICA	Aconcagua	6 959
NORTH AMERICA	McKinley	6 193
AFRICA	Kilimanjaro	5 895
ANTARCTICA	Vinson	5 140
AUSTRALIA	Kosciusko	2 230

OCEAN	TRENCH	DEPTH (m)
PACIFIC	Philippines	11 516
	Mariana	11 033
	Tonga	10 882
	(many others)	
ATLANTIC	Puerto Rico	9 200
INDIAN	Diamantina	8 047

Figure 2.4

3 The Earth's Composition

We have seen that there are three major realms of the planet Earth—the **atmosphere**, the oceans (**hydrosphere**), and that with the largest volume, the solid rock ball (**lithosphere**). We might add a fourth, the organic realm (**biosphere**). It is instructive to compare the compositions of these four realms, as set out in the chart below, Figure 2.5. Compare, for instance, the compositions of the ocean waters and living organisms; contrast the compositions of the atmosphere, the surface rocks and the oceans. Why should there be such a high proportion of oxygen in the surface rocks of the Earth, and none in the Core rocks?

REALM	Average depth (m)	Average relative density	COMPOSITION (% of atoms)					
			N	H	C	O	Si	Metals
ATMOSPHERE	300 000	0–0·0013	76	2		21		
OCEANS	2 630	1·04		66		33		
BIOSPHERE	0·1	1·0	1	60	8	30		
CRUST	17 000	2·8			3	60	20	16
MANTLE	2 883 000	4·5				54	14	29
CORE	3 469 000	10·71						100(?)

(CRUST, MANTLE, CORE grouped under LITHOSPHERE)

Figure 2.5

4 Other Earth Properties

There are just two which should be mentioned here. The Earth, as a ball of relatively dense rock, has a stronger **gravitational pull** than most of the other Solar System planets. This allows our planet to retain an atmosphere (in contrast to the Moon), and this property has further importance in relation to rainfall and the movement of ice and water across the surface: in short it is a basic factor in determining geological activity. We shall see later how slight variations in gravity measured at the Earth's surface enable us to understand the rock arrangements beneath the surface.

The Earth also possesses a **magnetic field**, as if there were a permanent bar magnet stretched from north to south approximately along the axis of rotation (ie from the North Pole to the South Pole). This fact has taken on an increasing significance in geological research in recent years, and we shall be referring to this in Chapter 14.

5 Earth History

Geological studies have shown that the Earth is extremely old: we speak in terms of hundreds and even thousands of millions of years. It is difficult to imagine such vast lengths of time. The chart, Figure 2.6, gives you a list of the periods of geological time, and some of the important dates involved. We shall be referring to these throughout the course, and in Chapter 15 you will find out how these periods have been divided up, named and dated.

A   B

Millions of Years ago	ROCKS CONTAIN PLENTIFUL FOSSILS	Era	PERIOD/Epoch	Length (Million years)	Characteristic Life
600		CAINOZOIC (Recent life)	QUATERNARY — Holocene	0·01	AGE OF MAN
			Pleistocene	2	
1000			Pliocene	9	
			Miocene	14	AGE OF MAMMALS
	FEW SIGNS OF LIFE		TERTIARY — Oligocene	15	
			Eocene	30	
		70 MILLION YEARS AGO			
2000		SECONDARY OR MESOZOIC (Middle life)	Cretaceous	65	AGE OF REPTILES, AMMONITES
			Jurassic	45	
			Triassic	40	
2700	Oldest British rocks	220 MILLION YEARS AGO			
3000	Oldest rocks in World	PRIMARY or PALAEOZOIC (Ancient life)	Permian	50	Amphibians, Fish, Insects, Coal Forests First Fish, Land Plants
			Carboniferous	80	
3900			Devonian	50	
4000	Moon rocks		Silurian	40	
			Ordovician	60	AGE OF INVERTEBRATES (No Land Life)
			Cambrian	100	
4500	Meteorite	600 MILLION YEARS AGO			
5000	ORIGIN OF EARTH?	PRECAMBRIAN		4000	Little evidence of life

Figure 2.6 This chart shows a time-scale of earth history on the left A and the main periods of geological time on the right B. What proportion of the earth's time-scale is occupied by rocks containing plentiful evidence of past life? The Cainozoic Era includes both the Tertiary and Quaternary.

THINGS TO DO AND DISCUSS

(1) Compare the details of the planet Earth described in this chapter with those estimated for other planets in the Solar System (eg size, gravity, composition, density, distance from the Sun, length of day, etc). Compile your own chart to summarise the comparisons.

(2) Copy out the chart of geological time (Figure 2.6B) in the front of your notebook so that you will always have it available for reference. Draw a 'Time Line' to scale, and mark in some of the major events of Earth history as shown on Figure 2.6A and Figure 2.6B. Emphasise the length of Precambrian time.

3 Evidence from the Surface Rocks

The study of geology has been likened to a detective story. The geologist finds his clues in the rocks which have an outcrop at the Earth's surface, and from his studies of the evidence provided there he is able to suggest solutions to the puzzling mystery of what happened in the past, long before man recorded anything in written form.

You have probably been struck by the variety of rocks you see in the cliffs at the seaside, or in the museums you have visited. They are of different colours and grain-sizes; some are extremely hard and compact, others soft and crumbly; some contain fossils and others do not; and there are a variety of structures criss-crossing the rocks. All these features are significant when we are attempting to work out the past history of the Earth, and it is important for us to be trained to observe them efficiently, so that our records are intelligible to other geologists.

A First Look At Rocks

At this stage we shall be concerned only with looking at some of the main characteristics exhibited by rocks: later in the course the various groups will be examined in greater detail. A number of introductory investigations are suggested at the end of this chapter.

When you handle a piece of rock with the aim of making a geological investigation it is important for you to make some definite observations of the features it shows. These should include some actual measurements and should use precise terms.

(1) **A rock is composed of various particles.** Can you distinguish them, or are they so small that the rock appears to be formed only of one type of material? A useful distinction can be made between rocks where the particles are easily visible with the naked eye (coarse-grained), those which have a rough feel to the touch and whose particles can only just be distinguished (medium-grained) and those which are smooth (fine-grained). Use rocks in your school collection to identify these types, and then measure the sizes of the grains in each type of rock. Where would you place the division between fine, medium and coarse?

(2) **If you cannot see the individual particles clearly** (ie if the rocks are medium- or fine-grained) there are other questions you can ask about the rock specimen.

(a) What is the **colour** of the whole rock? Certain rocks have distinctive colours: chalk (sedimentary rock) is white; basalt lava (igneous rock) and some fine shales (sedimentary rocks) or slates (metamorphic rocks) are black; limestone (sedimentary) may be grey or cream; rocks formed on the land containing iron oxide (sedimentary) are often red.

(b) Does the rock contain animal- or plant-like markings (ie **fossils**)? These are found only in the major group known as the sedimentary rocks, which are formed by the accumulation of layers of mud, sand, shell or plant debris. (Igneous rocks originate from the Earth's interior and solidify as they cool after being molten and fluid; metamorphic rocks have been subjected to great heat and pressure so that any fossils they once contained as sediments may have been distorted or destroyed.)

(c) What **small-scale structures** can be seen? If there are partings between the rock layers it is most likely a sediment (though igneous rocks are sometimes layered and metamorphic slates show fine cleavage structures).

(d) Is the rock a **carbonate**? Rocks containing a proportion of calcium carbonate are normally sedimentary rocks, and can be identified because they have a fizzy reaction with cold, dilute hydrochloric acid. If the calcium carbonate is the main component of the rock this is known as a limestone, but it may form only the cementing medium for other materials such as sand grains—when the rock would be known as a calcareous sandstone.

(3) **If you can see the individual particles** of which the rock is formed,

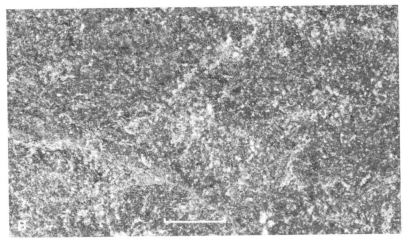

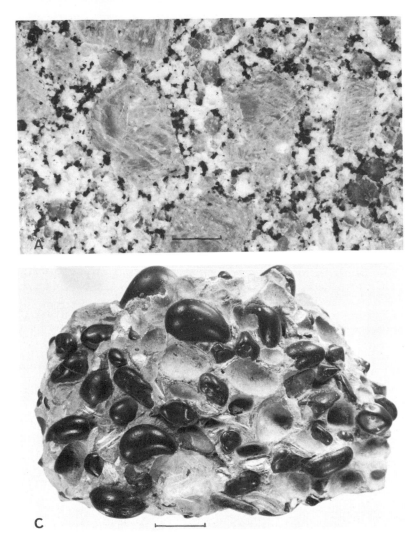

Four rocks. These photographs show how rocks are composed of minerals or rock fragments. The scale line represents 1 cm in each case.

Rock A is formed of interlocking crystals: the large grey crystals are pink potassium-rich feldspar (orthoclase); the smaller grey crystals are glassy quartz; the white crystals are sodium/calcium-rich feldspar (plagioclase); and the black crystals are mica. This rock is a granite, formed by cooling from molten rock material.

Rock B was also formed by cooling from a melt, but the crystals are much smaller than in rock A and include more dark minerals, like augite and hornblende. The lighter crystals are plagioclase feldspar. The rock is basalt.

Rock C is composed of rounded flint pebbles, cemented by a matrix of fine sand and silica. It is known as a conglomerate.

Rock D is formed of crystals of similar minerals to those occurring in rock A, but they have been squashed and drawn out in lines: this is typical of rocks formed under conditions of heat and pressure due to deep burial in the Earth's crust.

Describe these rocks in terms of the grain-sizes, minerals contained and other visible characteristics. Name them: igneous, metamorphic or sedimentary.

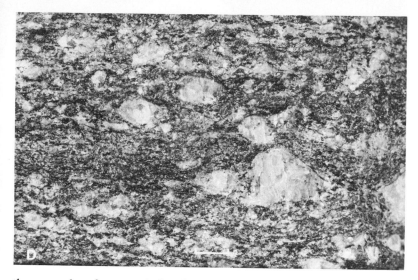

they may be of two varieties. Some may be rounded or angular rock-fragments (ie pebbles, sand-grains) set in a finer-grained cement: this will be a sedimentary rock. Or they may be a series of interlocking minerals, which formed from a mobile mass of molten rock material (ie igneous), or by recrystallisation under conditions of intense pressure and heat (ie metamorphic—these will often have the crystals drawn out in marked alignments).

These questions will have shown you that rocks have certain properties which can be observed, measured and recorded: each type of rock always has similar features. At the same time you will have noticed that it is not so simple a matter to divide a series of unconnected pieces of rock into even the major groups. Before you are able to identify and interpret the significance of the varieties of rock with confidence, you will need to understand more about a number of characteristics and processes.

(a) You should be able to identify the minerals which make up the rocks.

(b) You should understand something of the processes which lead to the formation of the rocks.

(c) It is also important to examine the associations of the different rock-types in their natural setting, and the distinctive patterns produced by their outcrops.

This is what we shall be doing in the rest of this chapter, and in Chapters 4 to 12. By the end of this part of the course you should be able to identify most of the common rock-types, and to use them as evidence concerning the geological conditions at the time of their formation.

LOOKING AT MINERALS

Rocks are composed of minerals, or of fragments of older rocks which are also made up of minerals. Sometimes we cannot see the minerals with the naked eye, and they can be identified only under the microscope. It is important for us to be able to recognise these minerals, because they tell us what type of rock we are examining and help us to begin to understand the rock's origin.

A geological definition of a mineral is 'an inorganic substance which occurs naturally and has a definite chemical composition and physical properties which vary within known limits'. Most minerals are compounds of several elements, and it is rare to find pure elements occurring naturally. The main elements combining to form minerals in the rocks of the earth's crust are oxygen (46·6 per cent of the total), silicon (27·7 per cent), aluminium, iron, calcium, sodium, potassium and magnesium: these eight make up 98·5 per cent by weight of all the rock-forming minerals.

Most minerals have a definite internal arrangement of atoms, which is occasionally expressed for us to see in the shape of a **crystal**. Crystals only develop their distinctive shapes where they can grow freely, and as minerals usually have to compete for space as they are solidifying these are rarely found. Crystals of one mineral always form similar geometric shapes, and the angles between corresponding faces

Crystal structure. On the right is a large single crystal of potassium alum: notice the eight-sided form (octahedron), and the way in which the smooth plane faces cut in sharp edges. The left-hand photograph shows a pile of polystyrene balls. What do these balls suggest about the arrangement of molecules within the upper half of the crystal, and the formation of the crystal faces?

of that crystal are always the same. Although we cannot see the internal atomic structure of a mineral, we can learn a lot about it by looking at the crystal form, which is a direct reflection of the way in which the atoms are packed together. Thus the sodium and chlorine atoms in common salt form a cube-shaped pattern, and the mineral crystallises in cubes. Some of the typical crystal shapes are illustrated on page 15. Crystals of one mineral may be joined together in many ways, but when a regular pattern is visible they are said to be **twinned**.

Twinning is very common in the feldspar minerals, and is an important diagnostic feature.

How Do We Identify Minerals?
Most of the common minerals are so distinctive that we can recognise them by just looking at specimens and noting various physical properties. For identification of some of the rarer minerals, and confirmation of the identity of the common ones, it is necessary to examine them

A

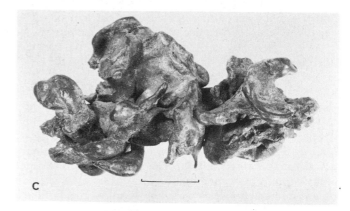

C

B

D

The scale line represents 1 cm in each case.

Minerals and crystals. Minerals occur as crystals showing faces related to the internal atomic structure only when free to grow in this way: normally other minerals are growing at the same time and prevent this.

A—Rock salt: a cubic crystal, cleaving (ie splitting) parallel to the face.

B—Fluorspar: two intergrown cubic crystals. Notice where cleavage takes place.

C—Native copper: it is unusual for an element to occur in the pure form. Does the shape of this specimen suggest how it might have formed?

D—Pyrite: intergrown crystals with pentagonal faces. These often have lined (striated) and pitted surfaces due to the imperfection of the natural processes of crystallisation and later chemical effects.

under a microscope or test them using a blowpipe and chemical reagents. It may even be necessary to study them by X-ray diffraction.

Fortunately all the minerals we shall be interested in can be identified without using special equipment, and the best way to do it is to ask ourselves a series of questions.

(1) What is the mineral's **colour**? This is often important, although impurities may be misleading. Quartz is usually colourless ('rock crystal'), but may be dark brown ('smoky'), rose pink, or mauve ('amethyst'). On the other hand iron pyrites (light, bronzy-yellow) and copper pyrites (brass-yellow) are distinguished by their colour.

(2) What is the **streak** like? The streak is more constant than colour, and is found by scratching the mineral on a piece of unglazed porcelain, by forming a fine powder in scraping the surface of the mineral, or by crushing it. The iron oxide, haematite, may be red or black, but its streak is always blood-red.

(3) What is its **lustre**? This is the appearance of the mineral in ordinary light. Is it shiny and metallic, shiny and brilliant like a diamond (adamantine), or shiny and glassy (vitreous)? Or is it dull? Is it opaque, or is it transparent?

(4) Has the mineral a **cleavage or fracture** that is distinctive? Many minerals break easily along planes of weakness in the crystalline structure. The resultant planes of cleavage form flat faces to the crystal: augite has two sets of cleavage planes (see photographs on pages 18–19). Other minerals have no cleavage and may fracture in an irregular but distinctive way. Quartz, for instance, breaks with a conchoidal ('shell-like') pattern, asbestos with a fibrous, thread-like fracture.

(5) How **hard** is the mineral? Minerals vary in their resistance to scratching, and we use Mohs' scale to compare them:

1 Talc (very soft)	6 Orthoclase feldspar
2 Gypsum	7 Quartz
3 Calcite	8 Topaz
4 Fluorite	9 Corundum
5 Apatite	10 Diamond

None of us is able to carry a set of these minerals around with us, but a strong finger-nail, copper coin and pocket-knife are often useful. Talc and gypsum can both be scratched by a finger-nail ($H = 2-2\frac{1}{2}$), and calcite is roughly equal in hardness to a copper coin. One of the most important distinctions to make is that between quartz and calcite, which both occur as white veins in rocks: a pocket-knife with a hardness of about $5\frac{1}{2}$ between apatite and orthoclase, will scratch calcite easily but it will not affect quartz.

(6) Is the mineral **heavier** than similar-sized specimens of other minerals? This will reflect the densities of the minerals compared: lumps of many metallic ores such as those of lead or iron are quickly recognised by their weight when compared with pieces of ordinary rock or non-metallic minerals.

(7) What **habit** does the mineral have? This is a comment on the mineral's shape. Does it show crystal faces, or does it have a rounded, nodular shape, or is it without any particular shape (massive)?

(8) Has the mineral any other distinguishing properties? Some may be **magnetic** (iron oxides like magnetite), and some may be **easily dissolved** in weak acid. Calcite can also be distinguished from quartz because it fizzes in weak hydrochloric acid, whereas quartz does not react.

All these properties, which we shall be using to identify the various minerals, are directly related to the internal atomic structure of the minerals: the forces between the atoms and the way in which they are 'packed' determines the cleavage and hardness, and the atomic weights of the constituent elements affect the density.

The Polarising Microscope

The polarising microscope is a very important instrument which geologists use to help in the task of identifying minerals and examining rocks. It is often called the Petrological Microscope because of its use in the study of rocks (ie Petrology). In order to make a section of rock thin enough for light to pass through a thin slice has to be cut with a diamond saw; one face of it is polished and fixed with Canada Balsam to a glass slide; the other side of the rock slice is then ground down and polished until it is 0.03 mm thick.

A simple type of polarising microscope is drawn in Figure 3.1 to show the main features. It is just like the ordinary microscope you may have used in Biology, but has two pieces of **Polaroid**, which have the property of polarising the light passing through them. Ordinary light vibrates in all directions in a plane at right angles to the direction of travel, but polarised light only vibrates in one direction. One piece, the polariser, is below the stage; the other, the analyser, is above it.

When we use the polarising microscope we are able to examine a mineral in three different ways, during each of which we can record certain diagnostic features of the mineral as the light passing through it is modified.

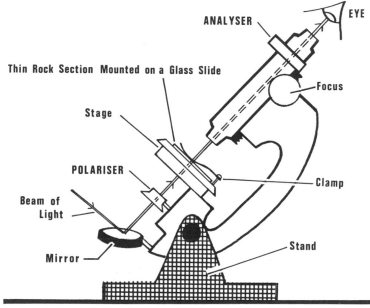

Figure 3.1 A polarising microscope.

First, we can take the Polariser and the Analyser out of the beam of light passing up the microscope, and examine the rock section as ordinary light passes through it into our eye at the top. We can note (a) that most minerals are transparent and colourless, but that a few are opaque and coloured; (b) that some minerals stand out from the others and have a high relief with strongly marked outlines and even a pitted surface; (c) that the first minerals to form have a good crystalline shape, and it may be possible to refer them to a definite crystal system; (d) the cleavage or fracture present; and (e) if fossils are present and attempt to identify them.

The Polariser can then be inserted into the beam of light, and the mineral examined as plain polarised light (**PPL**) passes through it. The main property to be noticed is whether the coloured minerals change colour as the stage rotates. Some do so markedly, and are known as **pleochroic** minerals. Pleochroic minerals show different colours according to the orientation of their crystals with respect to the direction of vibration of the polarised light. The best example is biotite, which changes from brown to yellow as the rock section is rotated on the microscope stage.

The first two methods of examination are usually combined when one is using a polarising microscope, because the facts which we have suggested should be noted under ordinary light can also be seen under plain polarised light.

When both the Polariser and Analyser are inserted, the pieces of polaroid are crossed (**XP**). We note that (a) a few minerals, mostly of the cubic system, are black however the stage is rotated; (b) most of the minerals have distinctive colours varying from bright to pastel shades and from greys to reds, greens and blues; and (c) some minerals show evidence of twinning, where half the mineral is extinguished when the other half is lit up and vice versa.

When you are examining a thin section of rock in this way it is best to recognise each mineral contained in the rock, noting down all the distinctive features which enable you to identify it. Then compare it with the other minerals in the slide. Is it bigger than the rest? Has it a better shape? What proportion of the slide does it occupy? When you

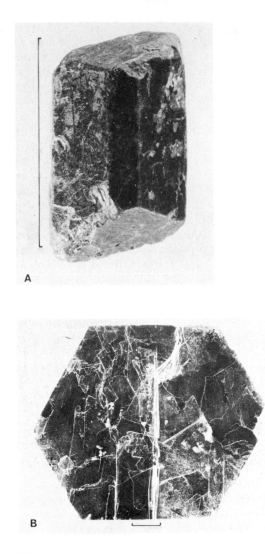

A

B

C

D

18

E

F

G

Rock-forming minerals. NB These perfect crystals seldom occur in rocks.
A—Augite; B—Biotite mica; C—Potassium-rich feldspar, orthoclase;
D—Quartz; E—Garnet; F—Calcite, showing rhomb-like cleavage;
G—Calcite, 'dog-tooth spar'. The scale line represents 1 cm.

19

have treated the other minerals in a similar way, make simple drawings of representative parts of the slide, and finally suggest a name for the rock from which it was cut.

THE COMMON ROCK-FORMING MINERALS

Most of this chapter so far has given a general introduction to the study of minerals. Three groups of minerals interest geologists, the gemstones, the ores of economic value and the minerals which are important in making up the rocks. We shall come back to the first two groups in Chapter 10, and only the last is to be dealt with here.

We have already noted that oxygen and silicon are the two most important elements in the rocks of the Earth's crust. They combine to form silica (SiO_2), the commonest form of which is quartz. The great majority of the minerals found in the crustal rocks are silicates, which may be thought of as combinations of silica with the oxides of other elements. This is particularly true in the igneous and metamorphic groups of rocks, where they are absolutely dominant. The **igneous rocks**, which include all those which were once molten, have seven 'families' of minerals which make up 99 per cent of all the rocks in the group. They are the olivines, pyroxenes, amphiboles, micas, feldspars, feldspathoids and quartz. The **metamorphic rocks**, which have been altered in various ways by heat and/or pressure from their original nature, have a wider range of minerals, but they are still mostly silicates. The third major group, the **sedimentary rocks**, are formed by the accumulation of layers of rock debris such as sand and mud, or by the piling up of shells, decaying vegetation or chemical deposits. The minerals contained in this group include some 'inherited' from the other groups, like the more resistant quartz, and some newly formed by chemical reactions including the clay minerals (also complex silicates), the carbonates and evaporated salts.

All the common rock-forming minerals can be classified in these groups:

THE SILICATE MINERALS

The Olivine group
The Pyroxene group, eg augite
The Amphibole group, eg hornblende
The Micas and Clay Minerals
The Feldspar group
The Feldspathoid group
The Quartz group
The Garnet group

THE NON-SILICATE MINERALS

The Carbonates
The Halides, eg rock salt
The Sulphates, eg gypsum
The Oxides, eg magnetite
The Sulphides, eg pyrite

A major division we recognise amongst the silicate minerals is between the dark-coloured group containing a high proportion of iron and magnesium—known as the **ferromagnesian** minerals—and the lighter-coloured feldspars and quartz—known as **felsic**. Figure 3.2 summarises the diagnostic properties of the main rock-forming and ore minerals.

THE SILICATE MINERALS

The Olivine Group
These have the simplest chemical formula and the smallest proportion of silica, and this is why they are mostly found in dark-coloured igneous rocks in which there is less than 55 per cent silica. Most olivines are a mixture of iron and magnesium with silica: $(Mg,Fe)_2SiO_4$. The atomic structure is tightly bonded, and olivine forms compact crystals having a glassy colourless appearance. Olivine crystallises at very high temperatures, and therefore often has a good crystalline shape, but it is easily altered to the green mineral serpentine and black iron oxide. When examined under the microscope (PPL) olivine is colourless, but has a high relief with hard, black outlines and a pitted surface; it is crossed by arc-like fractures, and the greenish alteration products can often be seen eating into the margins of the mineral. When the polars are crossed (XP) olivine has bright polarisation colours varying from reds to blues and greens.

The Pyroxene Group
Augite is the most common member of this group, which has a higher proportion of silica than the olivines. The pyroxenes are a large group, also found in the dark igneous rocks, and augite itself may vary in its composition. Its formula is $(Ca,Mg,Fe,Al)_2(Al,Si)_2O_6$, with atoms of aluminium replacing some of the silicon. Augite forms 4- or 8-sided crystals and the cleavages are a feature of the mineral when seen under the microscope. It is usually colourless or pale brown (PPL) with bright polarisation colours (XP).

The Amphibole Group
Hornblende is the best-known representative of this group which has more silica and is an even larger group than the pyroxenes. One feature of the amphiboles is their long, fibrous crystals and many of the asbestos minerals occur amongst them. The composition of these minerals is very complex, but similar to the pyroxenes with the addition of hydroxyl (OH): they can be termed hydrous calcium-magnesium-iron silicates. Hornblende forms long, flat crystals which are dark in colour like augite, but which have a glittering, silky sheen in contrast to the dull surface of augite. In thin section it is usually pleochroic from green to yellow, with marked cleavage lines (PPL) and long crystals, but the polarisation colours (XP) are often obscured by the body colour of the mineral.

The Micas and Clay Minerals
The minerals in this group are hydrous potassium–aluminium silicates, and the most distinctive feature of the micas is their flaky nature due to the cleavage planes along weak bonds between the strong sheets of silicate molecules. There are two main varieties: **muscovite**, which is lighter in colour, and **biotite**, which is darker because it contains iron. Both have the typical strong cleavage which gives ragged outlines to the crystals examined under the microscope. Muscovite is colourless (PPL) but bright pink and green (XP), whereas biotite is dark brown in colour (PPL) and very pleochroic (to pale yellow). Biotite does not change much under XP, as the strong body colour swamps the pinks and greens.

The clay minerals are related to the micas, and are formed when feldspars and ferromagnesian minerals break down. They are hydrous aluminium silicates, which are easily transported by rivers to form layers of mud. The individual grains are so small that they can be distinguished only under a very powerful microscope. **Kaolinite** is one of the most important, and is the main constituent of China Clay.

The Feldspar Group
These anhydrous potassium, sodium and calcium alumino-silicates are the most important of all the minerals occurring in the igneous rocks and may compose as much as 75 per cent of a granite. The two main varieties are **orthoclase** (potassium-rich) and **plagioclase**, which is a whole series of minerals with varying proportions of sodium and calcium. The sodium-rich varieties commonly occur with orthoclase in the light-coloured igneous rocks, whilst the calcium-rich plagioclases

Mineral	Composition	Colour	Hardness	Density	Streak
I. ROCK-FORMING MINERALS					
OLIVINE	$(Mg,Fe)_2SiO_4$	Colourless	6–7	3·2–4·3	
AUGITE (Pyroxene)	Silicate of Ca,Mg,Fe,Al	Black	5–6	3·2–3·5	
HORNBLENDE (Amphibole).	Silicate of Ca,Mg,Fe,Al,Na	Black	5–6	3–3·47	
BIOTITE MICA	Silicate of Mg,Al,K,H,Fe	Brown	2·5–3	2·7–3·1	
MUSCOVITE MICA	Silicate of Al,K,H	White	2–2·5	2·76–3	
ORTHOCLASE FELDSPAR	$KAlSi_3O_8$	White, pink	6	2·57	
PLAGIOCLASE FELDSPAR	$(Ca,Na)AlSi_3O_8$	White, grey	6	2·6–2·76	
QUARTZ	SiO_2	White, colourless	7	2·65	
CALCITE	$CaCO_3$	White	3	2·71	
GARNET	see Fig. 10.16				
II. METALLIC MINERAL ORES					
MAGNETITE	Fe_3O_4	Black	5·5–6·5	5	Black
HAEMATITE	Fe_2O_3	Red/black 'skin'	5·5–6·5	4·9–5·3	Cherry red
LIMONITE	$2Fe_2O_3.3H_2O$	Bronze/yellow	5–5·5	3·6–4	Yellow-brown
PYRITE	FeS_2	Pale bronze-yellow	6·5	4·7–5·1	Greenish-black
COPPER PYRITE (Chalcopyrite)	$CuFeS_2$	Brass yellow	3·5–4	4·1–4·3	Greenish-black
MALACHITE	$Cu_2CO_3(OH)_2$	Bright green	3·5–4	4	
AZURITE	$Cu_3(CO_3)_2(OH)_2$	Azure blue	3·5–4	3·7–3·8	
GALENA	PbS	Lead grey	2·5	7·5	Lead grey
ZINCBLENDE (Sphalerite)	ZnS	Black/brown	3·5–4	4	White
CASSITERITE	SnO_2	Black + pink powder	6–7	7	
III. OTHER MINERALS					
FLUORSPAR (Fluorite)	CaF_2	Usually purple	4	3–3·25	
BARITE	$BaSO_4$	White/pink/blue	3–3·5	4·5 ('Heavy spar')	
GRAPHITE	C	Black/dark grey	1	2–2·3	Black
ROCK SALT (Halite)	$NaCl$	Colourless (pure)	2·5	2·2	
GYPSUM	$CaSO_4.2H_2O$	Colourless (pure)	2	2·3	

Figure 3.2

Other Properties	Commonest Modes of Occurrence	Importance, Uses
Green alteration	Basic igneous rocks	
Prismatic crystals	Basic igneous rocks	
Prismatic crystals	Igneous and metamorphic rocks	Asbestos is one form
Thin laminae	Igneous and metamorphic rocks	
Thin laminae	Many kinds of rocks	Electrical industry, lubricants
	Many kinds of rocks	
Multiple twinning	Basic igneous rocks	Pottery; glazes; mild abrasive
Vitreous lustre	Many kinds of rocks; gangue	
Reacts with cold, dilute acid	Limestones; gangue	Building sands; pottery; furnace-lining; electronics
Strongly magnetic, metallic lustre	Igneous, metamorphic rocks	Iron ore
	Veins, metamorphic rocks	Chief iron ore
Dull lustre	Weathered iron ores	Iron ore
Metallic lustre	Veins, nodules	Most common sulphide
	Veins	Copper ore
	Associated minerals formed by underground waters	Copper ores (e.g. Katanga)
Cubic crystals, metallic lustre	Veins	Chief lead ore
Adamantine lustre	Veins	Chief zinc ore
	Veins, placer gravels	Chief tin ore
Cubic crystals	Gangue; hydrothermal veins	Enamelling; steel flux
	Gangue; hydrothermal veins	White paint; paper
Metallic lustre	Veins; metamorphic rocks	Pencil 'leads'; electronics
Taste	Evaporite sediments	Chemical industry
Pearly lustre		Cement; plaster; paper

are found in the darker types. Orthoclase is white or pink in colour and has a strong cleavage which often exposes smooth, glistening surfaces. All feldspars are colourless under the microscope (PPL), and only have faint cleavage lines, but under XP have grey colours and can be distinguished by the patterns formed by twinning in the minerals: orthoclase crystals are sometimes found divided into two parts (simple twinning), but plagioclases almost invariably have a series of narrow parallel strips (multiple twinning).

The Feldspathoid Group
These form a less common group of minerals which are found in rocks with a small percentage of silica and high proportion of alkaline elements like sodium and potassium.

The Quartz Group
Pure silica is found in important quantities only in the light-coloured igneous rocks, eg granite. The original molten material contained so much silica that some was left over when the other substances had combined with as much as they could, the residue then crystallising on its own. When seen in granite it forms colourless, rather glassy looking (vitreous) crystals or white veins. Quartz is one of the hardest minerals, harder than steel, and very resistant to chemical action: it is often preserved as rounded grains in sedimentary rocks. When examined under the microscope quartz is recognised by the properties it does not have! It has no colour, cleavage or fracture, and as it is usually the last mineral to form it fits in between the others without any recognisable crystal shape (PPL). The polarisation colours are greys, whites or yellows, unaffected by twinning patterns, although the wavy extinction pattern which is sometimes seen is evidence that the mineral has been subjected to stress.

The Garnet Group
Most of the minerals in Metamorphic rocks are silicates, but the more unusual members of groups like the pyroxenes and amphiboles tend to occur. The garnets are one of the most important distinct groups, forming dark red or green cubic crystals. Under the microscope they show up as large pink or grey crystals, but are always black under XP.

THE NON-SILICATE MINERALS

The Carbonates
Calcite and dolomite are the main non-silicates, and like the clay minerals are mostly found in sedimentary rocks and their metamorphosed equivalents like limestones and marbles. Calcite ($CaCO_3$) and dolomite ($MgCO_3.CaCO_3$) can be told apart because calcite effervesces with cold dilute acid, whereas dolomite needs warm acid.

The Halides and Sulphates
Rock salt ($NaCl$) and anhydrite ($CaSO_4$) are the most important minerals formed in evaporating lakes, and occur as soft crystalline deposits coloured by impurities. Rock salt is easily identified by its taste.

The Oxides and Sulphides
These are mainly iron minerals, like magnetite (Fe_3O_4) and iron pyrites (FeS_2). **Magnetite** occurs in small quantities in many igneous rocks, forming as black, diamond-shaped crystals. **Pyrite** is often found as crystals or nodules in sediments, where its silvery or bronzy-yellow colour is distinctive.

EARTH RESOURCES

The rocks and minerals occurring at the surface of the Earth are important not only as clues to the past history of our planet: they are the very basis of our everyday life! It is instructive to add up all the ways in which we as human beings are dependent on the resources of the Earth's crust. With what materials is your house built? From where do the materials come? Look around your home or school at the everyday objects you use: metal and plastic, powder, paste and paint all come from the rocks themselves, or from materials (eg oil) held in the pore spaces of the rocks; even the wood products and our food are ultimately dependent on the rock-derived minerals in the soil. The increasing world population and the depletion of these resources demands that we understand how they can be obtained and used to the best advantage.

The applications of geological knowledge are immense in an increasing variety of fields and examples will be discussed throughout this book. We are being made aware of the dangers of the wrong or selfish use of our environmental resources: it is false economy to rob nature by removing the mineral wealth from an area and not try to prepare that ground for another use.

THINGS TO DO AND DISCUSS

A practical notebook
Keep a special practical notebook as you work through this course for the investigations suggested at the end of each section. There is a lot of geology you can best find out for yourself, and you must train your powers of observation by regular practice.

Investigation 1. Rocks: a preliminary examination
The purpose of this exercise is to encourage you to look at rocks and to note some of the detailed features which they display.

Examine and describe six specimens of rock, which you have found yourself or which are supplied from the school collection. Write answers to the questions set out on page 11.

Investigation 2. Some common rock-forming minerals
Refer to the questions on page 16 and describe specimens of as many of the following common rock-forming minerals as possible: quartz, calcite, augite, hornblende, mica, feldspar, garnet, olivine. Hand specimens of the individual minerals should be compared with their occurrence in the rocks (eg quartz, feldspar and mica in a granite).

Construct a chart to summarise your description: you will not be able to complete every section for all the minerals. Write what you observe in blue ink, and what you have added from the chart, Figure 3.2, in red ink or pencil.

COMPOSITION	FORM	STREAK	COLOUR	HARDNESS	OTHER CHARAC-TERISTICS	NAME OF MINERAL

Investigation 3. Minerals under the microscope
(This will be possible only in a few schools where polarising microscopes are available, and it is more properly part of the Advanced Level course. We include this section, however, for those who wish to take their investigations a little further. If a polarising microscope is not available, the non-polarising type used in biology will show up many of the features seen under plain polarised light.)

Make coloured drawings of the rock-forming minerals listed in Investigation 2 as you see them in thin rock sections under the microscope (under plain polarised light (PPL), and under crossed-polars (XP)). This will give you at least two drawings for each mineral, and those which are pleochroic under PPL may need a third: these can be arranged neatly by drawing a cross on your page and inserting one mineral drawing in each of three sections. Compare the characteristics you notice under the microscope with those mentioned in connection with the mineral descriptions earlier in this chapter, and attempt to identify the minerals again in other thin sections. Note the rock-types in which they occur.

Investigation 4. Metallic ores and other minerals of economic value
Carry out the same instructions for Investigation 2 in the cases of some of the
following minerals (your list may be determined by the school collection or
examination syllabus): galena; haematite; pyrite; chalcopyrite; cassiterite;
halite (rock salt), fluorite; zincblende; limonite; magnetite; malachite; azurite;
barite; graphite; gypsum.

*Investigation 5. The applications of geological knowledge and the uses of
Earth resources*
Carry out the following investigations:

(*a*) Visit (after gaining permission) and describe any rock excavation being
made in your district. What type of rock is being quarried? What is it
to be used for? What methods of extraction and processing are in
operation?
(*b*) Examine the building and road materials used in your district. What are
they? Where is their source? Are they manufactured or processed in any
way, or do they come direct from a quarry?
(*c*) Notice the effects of geological processes on man's activities, and the
methods he uses to control the processes in your locality: eg coastal and river
defences, the building of dams, roads and bridges. Find out all you can
concerning the importance of geological knowledge in these projects.

Topics for discussion and essays
(1) Define the term 'mineral'. Explain with reference to actual examples, the
meaning of the following: cleavage, hardness, streak (W).
(2) Describe five different minerals and explain how they can be distinguished
from one another by the properties of streak, cleavage, lamellar habit, colour
and hardness (S).
(3) How would you distinguish, in hand specimens, between four of the
following pairs of minerals?

(*a*) Quartz and orthoclase feldspar	(*d*) Pyrite and chalcopyrite
(*b*) Olivine and augite	(*e*) Galena and zincblende
(*c*) Calcite and barite	(*f*) Fluorspar and rock salt (O).

Stob Dubh, Argyllshire. This was part of the glaciated Scottish Highlands.
The distinctive U-shaped valley was formed beneath the ice, but now the ice
has all melted, the rivers are attacking the rocks. (*Aerofilms*)

Part Two

Geological Processes Acting at the Surface Today

4 Weather and Wind

Most of the land surface of our planet is being worn down. The broken and dissolved rock debris produced in this process is largely transported to the sea and laid down on the sea bed. Periodically these deposits are uplifted to form new ranges of mountains and the process starts again. This is the dramatic series of events taking place at the Earth's surface, and is summarised on the opposite page.

We must study what is going on around us because we want to understand how our landforms have been produced; because we shall be able to understand more readily how the sedimentary group of rocks are formed from the debris that results; and because, as mentioned in Chapter 1, the basis of our geological studies is that 'the present is the key to the past'. We need to investigate what is happening now on each hillside, in the sea, in rivers and in glaciers, so that we can interpret the results of such action in the past.

This is going to be a most important part of our studies, but there are a number of points that must be borne in mind as we proceed.
(1) The **rocks** which outcrop at the surface have a controlling influence on the relief: they may be resistant or weak; they may cause rainwater to flow over their surface into valleys, or allow it to pass underground; or they may contain certain lines of weakness which will be easily etched out by the weather.
(2) Each landform—delta, cavern, sand-dune, cliff or corrie—is produced by **a distinctive process or agent** (eg running water in rivers or below the surface of the ground; the wind; ice in glaciers; the sea).
(3) A definite **sequence of landforms** is associated with the work of each agent, following the gradual wearing down of a landscape from an area of uplifted mountains or hills to an almost flat lowland. This sequence is often known as a **cycle**. It begins as a range of newly formed mountains is attacked by the atmosphere and its rocks begin to crumble (**weathering**). The **denudation**, or lowering of the whole landscape, is completed by various agents which act as cutting edges on the rocks (**erosion**), and **transport** and **deposit** the debris produced to form new layers of sediment. These become compacted and hardened (**lithification**) into solid sedimentary rocks, which are eventually raised up to form the next range of mountains and the cycle begins again.
(4) Each landscape is thus the product of the interaction of structure (rock hardness), process (river water, glacier ice, etc), and stage (in the cycle of landforms associated with each process). We must realise, however, that many parts of the world do not have a simple history which can be interpreted in terms of one process or of a single cycle. In Britain we find that river erosion is the dominant process, but that there are also many landforms which can only be attributed to a recent sheet of ice which covered most of the country. We also find that the level of the sea around our shores has not been constant, and where it has fallen there are the features of an earlier stage in the cycle of river erosion cut into an older valley: the river has cut down to a new level. Most of our landscapes thus have a **complex origin**, and every slope tells us something of the past history of their development.

We shall now investigate the various processes in detail, studying the ways in which they work to destroy, transport and to build up new relief features and layers of sediment. It is hoped that as you read this section you will become more aware of what is happening around you, and will take more notice of such events as rivers in flood, and waves breaking at the base of a cliff, as well as the results they produce. Unless you do you will never have a truly geological understanding of what the rocks tell us.

WEATHERING

As soon as rocks outcrop at the surface of the Earth they are attacked by the atmosphere, and reduced to rubble or soluble debris. There are two main types of mechanism at work, but they usually occur together, and their relative importance depends on the climate.

The Geological cycle. Mountains (*Aerofilms-Swissair*); Erosion and Transport (*Aerofilms-Swissair*); Folded Rocks (*Crown Copyright*); Deposition (*Aerofilms-Swissair*)

Physical Weathering

Changes of temperature are most important in regions where chemical reactions are limited, such as deserts and cold areas. In the hot deserts there are often great differences (up to 40°C) between the day and night temperatures. During the day the surface layers of rock are heated incessantly by the sun: all the minerals expand, but at different rates, thus creating stresses within the rock which tend to weaken it. Cooling is rapid at night, and people who have visited deserts have heard loud cracks as the rocks split apart under the continual strain of expansion and contraction, as shown in Figure 4.1. It has been shown that the presence of a small quantity of water (such as the dew forming at dawn) hastens this method of disintegration. The evaporation of water from rock surfaces allows the finer salts to be blown away, but the heavier iron salts are left as a coating of 'desert varnish'.

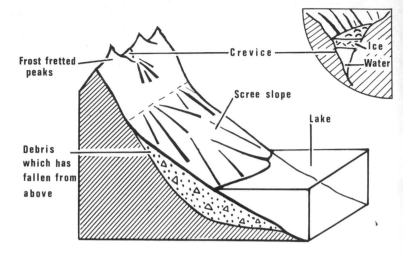

Figure 4.2 **Frost action on high peaks.** The enlarged inset crevice shows how the rocks are cracked: the surface ice freezes first, and when the underlying water freezes it will either force out the surface ice, or force apart the rocks.

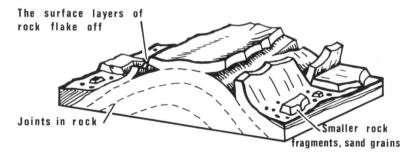

Figure 4.1 **Exfoliation, or onion-layer weathering.** Where would you find this type of rock-splitting?

Mountains and Tundra regions experience repeated freezing and thawing of water, since the temperature varies so much. The water collecting in pores, joints and crevices like the one in Figure 4.2 turns to ice overnight, and its volume increases by 9 per cent. Pressure is exerted on the surrounding rocks, and grains or larger fragments may be forced apart. The irregular, fretted appearance of the Alps and other high mountains results from this effect, and the tumbled masses of scree at the foot of slopes in such areas show us how active the process is.

It is interesting to notice the practical illustrations of these processes occurring around us. Quarrymen often heat resistant rocks to crack them apart; we see road surfaces broken after a severe winter; the continual fall of frost-shattered debris in mountain regions is a great problem in rail- and road-construction.

Rocks may also be broken as tree roots force them apart, as boring animals cut into them and even when lightning strikes a high peak.

Chemical Weathering

Rocks are decomposed most effectively in areas of warmth and high humidity. Our own British climate favours this type of

Wastwater, English Lake District. What has caused the formation of the fan-shaped masses of debris on the right-hand edge of the lake? (*Aerofilms*)

A dolerite sill which reaches the surface in Fifeshire. What is happening to the rock exposed to the atmosphere? (*Crown Copyright*)

weathering to a certain extent, but it is most rapid in the wet tropical areas.

The actual chemical reactions are very complex, and it is difficult to say which is most important. Rainwater absorbs some of the gases (oxygen, carbon dioxide), and collects minute salt particles as it passes through the atmosphere. It adds organic acids when it reaches the ground, and is thus able to start reactions in rock minerals which cause them to crumble, to be dissolved and carried away in the groundwater, or to change volume and exert pressure on the surrounding minerals.

Some minerals are not affected. Quartz, for instance, may go through several cycles of weathering, transport to the sea, deposition in a new rock and uplift in mountains, without being altered chemically. The feldspars break up more easily since the potassium, sodium and calcium are removed in solution as the remaining material is converted to clay minerals. Ferromagnesian minerals are least resistant and are also converted into clay minerals and soluble carbonates and oxides. Some of these reactions can be summarised as chemical equations, although the formulae of many of the minerals are very complicated.

$$6H_2O + CO_2 + 2KAlSi_3O_8 \rightarrow Al_2Si_2O_5(OH)_4 + 4SiO(OH_2) + K_2CO_3$$

| rainwater | orthoclase feldspar | clay mineral | silicic 'acid' | (in solution) |

Limestones, containing a high proportion of calcium carbonate, are dissolved by the dilute carbonic acid in rainwater: joints are widened to form **limestone pavements**, **solution hollows** are opened out and **caverns** enlarged underground. A special type of landscape, known as **karst**, results as demonstrated in Figure 4.3. The name karst comes

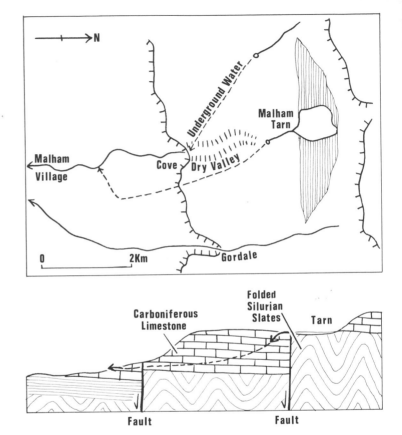

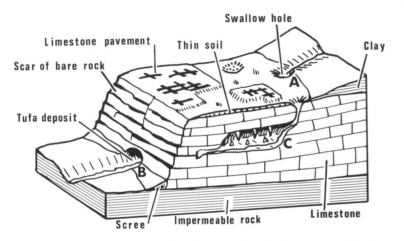

Figure 4.3 A karst landscape. There is no surface drainage across the limestone: a river disappears underground as it leaves the impermeable clay at A, and issues from the base of the limestone cliff at B. It flows through the cavern—C—which has a series of dripstone formations. What evidence of erosion and deposition can be seen?

Figure 4.4 The Malham area. Relate the geological structure shown on the section to the landscape and drainage features of the area.

from Yugoslavia, where limestone is a common rock. Such features can be seen in this country in the Mendip Hills, the Peak District of Derbyshire and the Malham–Ingleborough area of the central Pennines (Figure 4.4).

The weathering processes attacking a mineralised vein where it reaches the surface may lead to the solution and removal of the minerals at the surface and the enrichment of the lower part of the vein (Figure 4.5). Later erosion may then expose this richer section.

32

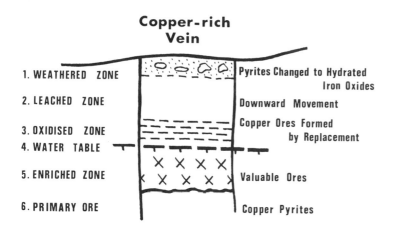

Copper-rich Vein

1. WEATHERED ZONE — Pyrites Changed to Hydrated Iron Oxides
2. LEACHED ZONE — Downward Movement
3. OXIDISED ZONE — Copper Ores Formed by Replacement
4. WATER TABLE
5. ENRICHED ZONE — Valuable Ores
6. PRIMARY ORE — Copper Pyrites

Figure 4.5 Enrichment by weathering. How does the change take place? Describe it in your own words.

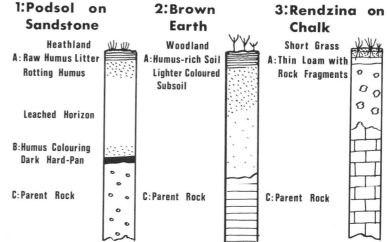

1: Podsol on Sandstone
Heathland
A: Raw Humus Litter Rotting Humus

Leached Horizon

B: Humus Colouring Dark Hard-Pan

C: Parent Rock

2: Brown Earth
Woodland
A: Humus-rich Soil Lighter Coloured Subsoil

C: Parent Rock

3: Rendzina on Chalk
Short Grass
A: Thin Loam with Rock Fragments

C: Parent Rock

Figure 4.6 Soil profiles. Try and account for the formation of the different soil layers. How far are the different types related to the rock-type? Podsol is a term originating in Russia, and Rendzina is used for thin limestone soils. Find out about other types of profile, including some of those near your home or school.

Soils

The weathered covering grows in thickness with the years unless it is transported away. In some tropical areas it is nearly 150 m thick, largely consisting of **laterite** and **bauxite** (hydroxides of iron and aluminium: bauxite is the main ore of aluminium), which become hard and encrusted near the surface. It seems that the hydroxides are concentrated near the surface by rising solutions in the dry season.

Other residual deposits, left after the solution of most of the underlying rock, include the **terra rossa** of karst areas, and the **clay-with-flints** on top of the Chalk Downs.

True soils are much more than the accumulation of weathered debris, and are only formed after a prolonged period of action by the plants growing on them, bacteria, worms, water and air. Their colour and layered arrangement are closely related to the climate, drainage conditions of the underlying 'parent' rock, and to the local relief. Thus heavy rainfall causes solution of the salts near the surface of the soil, and they are carried down (or leached) by the percolating water to the lower parts of the soil, which become enriched. In drier areas the salts are drawn up and concentrated near the surface. Mature soils are teeming with microscopic life, and on flat or gently sloping ground will be closely related to the climate of their area. Figure 4.6 shows sections through three contrasting British soils.

UNDERGROUND WATER

When rain reaches the Earth's surface it either runs off into the nearest streams, or percolates into the soil and rocks, or is evaporated.

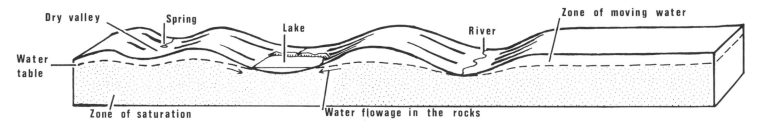

Figure 4.7 The water-table. How is the water-table related to the relief? What would be the effect of rainy and dry seasons on its position? Why should a spring rise where it is marked?

Percolation involves a small proportion of the rainwater, but it is very important: the amount depends on the underlying rocks.

(1) **A porous rock** is one with a large proportion of space not occupied by solid material, and where the grains are not tightly packed or cemented. Such a rock can hold a lot of water (an **aquifer**), or oil (a **reservoir rock**). Sandstones are often very porous.

(2) **A permeable rock** is one which will allow water to pass through it. It may be porous, or may be cut by wide joints. Thus a dune sand, which is very porous, is also permeable; a well-cemented limestone, cut by many joints, is non-porous but permeable; a clay may be very porous, but the spaces will be too tiny to allow water to pass from one to another, and is therefore impermeable.

The upper limit of rock saturated with water is known as the **water-table**, and the level tends to undulate in hilly areas, as demonstrated in Figure 4.7. Above it water is percolating downwards from the surface after a rainstorm, or may be moving upwards by capillary action in time of drought, thus forming a zone where a lot of solution and replacement of minerals takes place. A fossil shell composed of calcium carbonate may be dissolved away, leaving a hollow mould in the shape of the fossil, which may be filled by another mineral (like silica

or pyrite) at a later date. Movement is very slow below the water-table, and chemical deposition takes place, eventually filling the pore spaces and cementing the rock.

Concretions are formed as minerals distributed throughout a rock layer are concentrated into a single mass. Nodules of flint and pyrite are common concretions in the Chalk, and septarian nodules of impure limestone veined with calcite are found in the London Clay.

Springs occur where the groundwater reaches the surface, as shown in Figure 4.8, and they may have encrustations of various minerals around them. **Tufa** formed in this way is common in limestone areas (Figure 4.3). **Dripstones** are formed where the water percolating through limestone enters the roof of a cavern hollowed out of the rock. This water is a solution of calcium bicarbonate, from which calcium carbonate (calcite) crystallises and is left behind on the ceiling of the cavern to form a **stalactite**, and part of which drips to the floor to build up a **stalagmite**. The caverns themselves (Figure 4.3) and their connecting galleries are formed by a combination of solution and river erosion by underground streams. These streams are fed by rivers diverted underground through vertical solution pipes known as **swallow holes** or sinks (or dolines in Yugoslavia). Gaping Ghyll to the

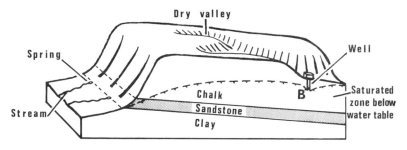

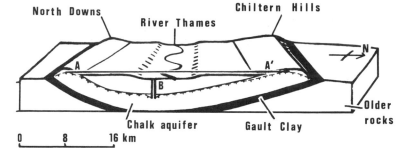

Figure 4.8 Springs and wells. Springs and wells in a Chalk area (eg North Downs). Which of the rocks shown is permeable and which impermeable? A well sunk into the Chalk will create a cone of depression at B.

Figure 4.9 An artesian basin. The London Basin (a very generalised section). The water-table level (A-A′) is above the top of the well at B, where the water will gush out under pressure.

south-east of Ingleborough in the Pennines is 120 m deep and leads down to a cavern. The caverns are gradually enlarged until the roof collapses, leaving an irregular depression or valley.

An understanding of what happens to water in the rocks underground is important in relation to the supply of water to rivers, the continuing solution and cementation of the rocks themselves, and also to water supply for human needs: water is one of the most important resources contained in the rocks. Man is using more and more water, and underground sources supply a small but important part of what we use in Britain. In the past local aquifers supplied a higher proportion, but many have been depleted by over-pumping. Thus the lower Thames valley around London is an artesian basin (Figure 4.9), but so many wells have been put down into the Chalk that the water-table has fallen. The fountains in Trafalgar Square were constructed to use the natural flow of water but now have to be supplied from pumps. Rivers and surface reservoir storage have become more important.

Areas with a smaller demand can still depend on local underground supplies. Thus the West Surrey Water Board extracts water from the

Hythe Beds of the Lower Greensand to supply Godalming and the surrounding villages. Figure 4.10 illustrates the geological situation. Compare this and that associated with the occurrence of oil (Figure 8.18).

There are problems connected with obtaining water from the rocks

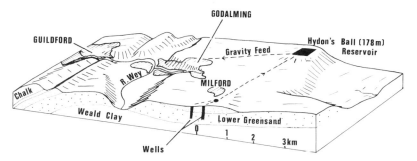

Figure 4.10 West Surrey water supply. Water is pumped from six wells near Milford to the Hydon's Ball reservoir, from which it is fed downhill to Godalming and district. The River Wey is also used as a source of water, and wells are dug into the Chalk near Guildford.

35

for either human consumption, or the growing of crops by irrigation —and crops are far more fussy! The water may become contaminated with salts, or even with sewage, and become unusable. Long, dry periods may lower the water-table to a level beneath the borehole tube. Many water boards in Britain obtain their supplies from rivers, or trap rainwater in large upland reservoirs, but the filtering processes are more expensive in such cases.

CREEP AND LANDSLIDES

Weathering may produce so much debris that it becomes unstable and moves downhill under gravity. Such migrations of rock waste may be rapid or slow, and may be 'triggered off' by an earthquake or sudden heavy rainstorm. The combined effect of weathering and mass debris movements of this sort is one of the most important processes wearing down the landscape.

The slower movements are less dramatic, but affect every slope and are more important. **Soil creep** is common in Britain, and you will have noticed the features illustrated in Figure 4.11, though the actual movement is imperceptible. Whilst the river is affecting a narrow zone at the bottom of its valley, soil creep is taking place over the rest of the landscape and assisting in the gradual lowering of the slopes.

In cold Tundra regions only the top layer of soil thaws in summer, and, since the melted water in the soil cannot sink through the permanently frozen layer underneath (the **permafrost** zone), the saturated surface soil flows slowly down-slope. This **solifluction** (soil-flowage) concentrates any fragments of rock in **stone stripes**, which become a distinctive feature of the hillsides. Even on flat ground the heaving motion of alternate freezing and thawing causes stones to be thrown out of the soil and to be piled up in **circular rings**. Such features are of common occurrence today in the very cold lands of northern Siberia, Alaska and northern Canada, but can also be found in areas which were once close to an ice sheet, like the British Isles.

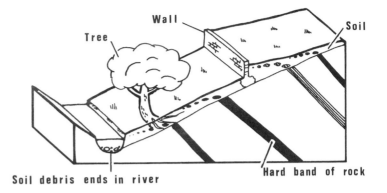

Figure 4.11 Soil creep. How has the downslope movement of soil affected the stones in it, the tree trunk and the level of soil on either side of the wall?

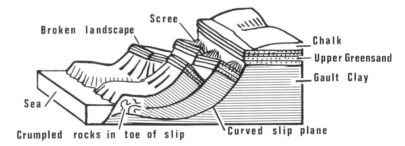

Figure 4.12 A landslip. Masses of rock have slipped forward along curved planes. This is due to the weak clay becoming saturated and collapsing under the weight of rigid overlying rocks.

The more rapid movements include **rock-falls**, **mudflows** and **landslips**. Mountainous regions, and those affected by sudden, heavy storms, are most liable to action of this type, which can often bring destruction and tragedy. Landslips are common in Britain where a rigid permeable rock lies on top of a weak, impermeable layer. In

1927 a landslip like that illustrated in Figure 4.12 carried away the road around the southernmost tip of the Isle of Wight, and between 1968 and 1970 another 100 metres of road on the Blackgang Chine side has been carried away.

WIND ACTION

Most parts of the world are moist enough to support grasses and other plants which bind together the surface soils, but in deserts, semi-arid areas, and along coasts, the wind may blow away the finer particles of rock waste. It is important to note that the grasses have only been in existence since the relatively recent Miocene period, and that wind action must have been much more important before that date. Nowadays it is only significant in the dry regions and although these cover a large part of the world's land surface, the results of wind action are puny when compared with the rapid and impressive results of river and glacial action.

Can the Wind Wear Away Rocks?
Winds vary considerably in strength from one moment of time to the next, and are thus only able to lift and move rock debris for short periods. At wind speeds of 40 km per hour sand grains (0·15–0·3 mm in diameter) will be moved, but extremely rare gusts of 150 km per hour are needed to roll pebbles along the desert surface. Fine dust (grains of less than 0·06 mm diameter) is moved most easily, and the lightest winds are sufficient to swirl it into the air. Dust storms, caused by temperature differences in the atmosphere, are common in the Sahara desert and the dust is carried far out over the Atlantic Ocean, and across Europe.

Sand grains are seldom raised far into the air, and are usually moved in surface-creep and saltation (by being bounced along). Whereas dust particles do not affect the rocks as they waft around them, sand may erode the lower few metres of a rock mass as it is thrown against it, and will also trim the larger pebbles in its path. The

results of such sand-blast action are seen in Figure 4.13. In addition the constant collision of sand grains in the air and on the ground makes them rounded, and gives them a frosted, **'millet-seed'** appearance.

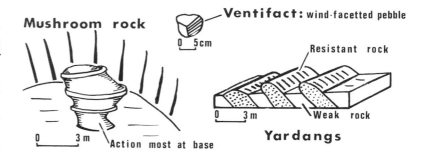

Figure 4.13 Sand-blast action. Notice the small size of the features produced.

Wind only erodes when it is carrying sand, and then only produces minor effects. It takes long ages to produce any noticeable result because it is active for such short periods, and only uses the tiniest grains for battering against hard rocks. The relative speed of weathering and wind action in desert areas and humid regions can be illustrated by Cleopatra's Needle, which stood for thousands of years in Egypt with no deterioration of the carved inscriptions. It has now been on the Thames Embankment in London for 100 years, and the detail of the writings has been destroyed by the corrosive atmosphere.

The most characteristic features of deserts are the widespread areas of bare rock and immovable broken blocks. Any fine dust is soon removed by the wind, and the sand rolled into local concentrations. Only one-eighth of the Sahara desert is covered by sand.

The wind also lowers the landscape by removing large volumes of

loose, dry sand and dust to leave wide hollows. This process is known as **deflation**. It is common in semi-arid areas where the protective grasses and trees have been removed in the process of farming, and the 'Dust Bowl' area of the Mississippi basin, and some of the 'virgin lands' of the USSR, have suffered the removal of vast quantities of dry soil in recent years. There are a series of large hollows in the Egyptian desert west of Cairo which were also caused in this way: the Qattara Depression, 135 m below sea-level, is the largest, and has its floor where the water-table reaches the surface.

Wind Deposits

As the wind slackens and drops, the sand, and eventually the dust, fall back to earth, forming distinctive deposits on the land, and adding to the marine sediments far out into the oceans.

Dust often settles far from the desert of its origin, and becomes anchored by vegetation. The **loess** of northern China covers an area the size of France. It consists of a fine-grained, yellowish loam, becoming coarser towards the Gobi desert, its probable source. Similar deposits are found in the Sudanese 'cotton soils', and round the margins of the former ice sheets in Europe and North America. In north Germany and the Netherlands earlier moraines stood out in front of the last ice sheets, and their bare, frozen surface was crossed by strong winds blowing down from the ice. All the fine debris was winnowed away and redeposited in northern France (the limon deposits) and central Germany (loess), leaving infertile, gravelly heathlands behind.

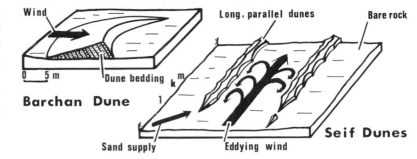

Figure 4.14 Sand dunes. Make a note of the internal structure of the crescent-shaped barchan (dune-bedding).

Sand dunes are temporary accumulations of wind-blown sand, and may be confused heaps, sand seas, or the more regular **barchans** and **seifs**, both of which are illustrated in Figure 4.14. Dunes are also common at the back of sandy beaches, and have resulted from on-shore winds blowing over the dry, bare beach at low tide. The dunes tend to move inland, and are only halted by planting them with special grasses and pine trees. Small areas of these coastal dunes are common around British coasts, including the Braunton Burrows in north Devon, and the Culbin Sands in Morayshire.

Dune sands. These structures are found in the Triassic rocks of Cheshire. How can we tell they were formed in desert conditions? (*Crown Copyright*)

5 Running Water

Running water plays the dominant part in wearing down our landscape today. Most of the world has enough rain to support perennial rivers, and the tiny amount of rain falling in deserts is important, since it falls in such heavy showers and the wind acts so feebly. River water is extremely effective in wearing away the land and in transporting rock debris to the sea: so much so, that the River Mississippi, for instance, carries 2 million tonnes of silt into the sea each day.

How Do Rivers Begin?

Sheets of rainwater spread over a hillside may cause the rapid removal of a lot of soil in the tropics, where up to 100 tonnes of soil per acre may be lost in a heavy storm, but most of this water is soon guided by the undulations of the ground into rills. Eventually a drainage pattern is built up, based on the slope of the land, and concentrating the run-off of water into a major stream flowing towards the sea. Rivers are not, of course, supplied with water only after a rainstorm, but are kept going by spring water, which contains dissolved salts gathered underground, and by water slowly released from peat, soil and plants.

How Does a River Erode?

A river having formed its bed, and been provided with water, flows at a speed determined by the slope, or gradient, of its course and the amount of water supplied to it. We have all noticed the rapid flow of mountain streams and the more lethargic movement of wide rivers on a flat lowland, and compared the normal sluggish flow with the rapid spate following a heavy storm. The sides and bed of a river cause the flow to be slowed down by friction, and the water moves most rapidly where it is deepest. Rivers flowing past banks of weathered debris will easily undercut large sections and carry them away, especially when

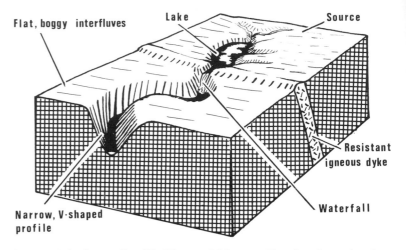

Figure 5.1 A river valley (1): The youthful stage. The river is cutting down into the rocks forming the original relief. Outcrops of hard rock protrude into the stream bed to form waterfalls (eg Glenashdale Falls, Arran, and Hardrow Force in Wensleydale, Yorkshire, illustrated in Figure 5.4). Lakes form in the original hollows along the course. All these 'irregularities' are gradually removed as the hard rock is worn down, and the lakes are filled in. Much of Britain's highland country is in the youthful stage.

flowing turbulently after a storm. They will also use the debris they are carrying to grind away (or **corrade**) the rocks of the river bed, and to wear down the boulders lying in the stream. Stones moved long distances by a stream are noticeably rounded. Swirling stones will eat out circular **pot-holes** (not to be confused with those found in limestone areas), the coalescence of which lowers the river bed. Pictures of the Lynmouth flood disaster of 1952 show us just how powerful a river can be, both in the size of boulders it can carry and in the damage it can cause. **Solution** may also take place on the river bed, especially in limestone regions.

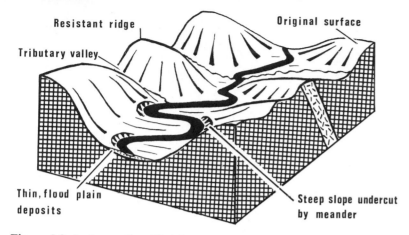

Figure 5.2 A river valley (2): The mature stage. Most of the irregularities have been removed from the river's course, and it has cut down to a graded profile. This means that it flows just fast enough down the slope it has cut to remove all the debris supplied to it by soil-creep on the valley sides. There may be a smooth general decrease in the river gradient from source to mouth, but there are often local breaks with steeper sections near a tributary which adds a large quantity of gravel to the load. Erosional energy is devoted to opening out the valley by undercutting the sidewalls. Meanders and a narrow flood-plain develop. Tributaries increase the size of their valleys in the same way and become very important in weak rocks: the valleys there may be wider than the main one. The main original stream is known as the con-sequent, and its tributaries as subsequent streams.

Figure 5.3 A river valley (3): Old age. A landscape of gentle slopes and low hills. The landscape cannot be worn down below sea-level, which is the base-level of river erosion, and the river has to maintain a gentle gradient to keep it flowing. Low hills (50–75 m above the valley floor) separate one valley from the next drainage system, but these residual masses, known as monadnocks, occupy a very small area. Such areas of low relief are known as peneplains, although few can be found in the world at the present moment, since a period of mountain-building has taken place so recently, followed by frequent changes in sea-level. Local examples can be seen on weak clay rocks. The river follows an intricately-meandering course, characterised by a shifting channel, ox-bow lake and cut-offs. These features are shown in the section of the river Mississippi drawn in Figure 5.5.

How Are Valleys Formed?

River valleys are formed by the combined action of the river cutting down through the underlying rocks, and the weathering and mass-wasting of the walls on either side. **Canyons**, with very steep sides, are formed in dry conditions, where there is not enough water to encourage weathering and soil creep. Valleys in Britain, however, develop different forms as time advances, and we recognise a progression of landforms, as illustrated in the block diagrams of Figures 5.1–5.3.

A V-shaped valley: Ashes Hollow, in the Longmynd, Shropshire. Which stage has river erosion reached here? (*Crown Copyright*)

River Patterns

The description of a river valley and its evolution as shown in Figures 5.1–5.3 is of course a very general picture, and local details vary according to the nature of the underlying rocks. The pattern of rivers and their valleys belonging to one drainage system will be based at first on the original relief of the area, but during the mature stage the streams will begin to emphasise the lines of weakness in the rocks. Some of these patterns are drawn in Figure 5.6: note which are guided by the relief, and which by weaknesses and structures in the rocks.

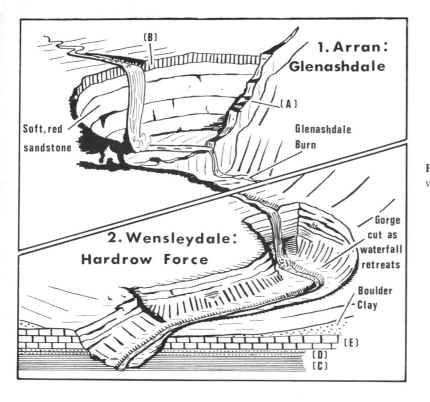

1. Arran: Glenashdale

(B)

(A)

Glenashdale Burn

Soft, red sandstone

2. Wensleydale: Hardrow Force

Gorge cut as waterfall retreats

Boulder Clay

(E)
(D)
(C)

Figure 5.4 Waterfalls. Glenashdale Falls are caused by two bands of hard igneous rock, a vertical dolerite dyke at A and an almost horizontal sill of felsite and dolerite at B. They have been intruded into softer sandstone.

Hardrow Force lies at the head of a 400 m gorge, which has been cut as the waterfall has worked its way upstream through a typical succession of Yoredale Series rocks. The weak shale (C) is removed easily, undermining the hard sandstone (D) and limestone (E) above. This fall, near Hawes in Wensleydale, is the highest in England.

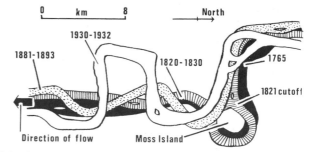

0   km   8   North

1930-1932

1881-1893

1820-1830

1765

1821 cutoff

Direction of flow   Moss Island

Figure 5.5 The Mississippi's shifting course. A comparison of various surveys made between 1765 and 1932.

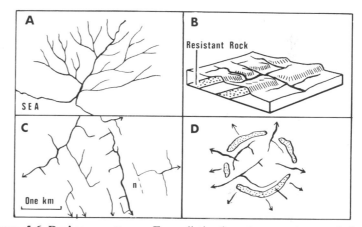

A

SEA

B

Resistant Rock

C

One km

n

D

Figure 5.6 Drainage patterns. Four distinctive stream patterns: A has a dendritic, tree-like arrangement, and is often found where there is little difference in the underlying rock-types; B is a trellis pattern, and D is radial, both being the result of alternate layers of hard and soft rock which have been tilted (B) or updomed (D); C shows the rectangular pattern found on Bodmin Moor, formed as lines of weakness within the granite rock have been etched out. Make a point of examining maps and noting examples of the different patterns.

43

River Capture

Some of these regular patterns are modified when the water flowing along one stream is diverted into another. This occurs when the headwaters of one stream cut back and erode into another valley flowing across the course of the first stream. If these headwaters are flowing at a lower height the water from the attacked river will be captured, and will flow into the drainage of the pirate stream, as shown on Figure 5.7. The erosional power of this stream will then be increased, and the process tends to continue once it has begun, leaving a series of beheaded streams which receive insufficient water to erode their valleys in the way they previously did. They become **underfit** streams. The history of the Rivers Wey and Blackwater, tributaries of the River Thames in Surrey, illustrates this process, and can be followed in Figure 5.8. Look at maps of the Yorkshire rivers draining to the Humber estuary, and of the upper Tyne system, and see if you can tell where river capture has taken place.

Another type of river capture takes place on a smaller scale where two sets of headwaters are competing against each other, or where one stream has a short, steep course to the sea and another has a long, gentle gradient. Figure 5.9 shows how both of these cases are to be found in the Cotswold Hills.

Figure 5.7 River capture (1). The waters of stream X are diverted to join those of stream Y, which has a steeper course and greater erosional power. One of the tributaries of Y cuts backwards into the X valley, and captures the headwaters. This leaves a section of the X valley without any drainage, and in places (Z) the stream flow may be reversed. Look at a map of the area immediately inland from Bude in north Cornwall, and compare the rivers draining to the sea at Bude with the river Tamar. Is river capture likely to happen here?

A: The Origin of the Drainage Pattern

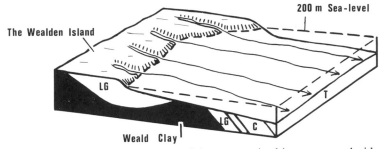

A. Shows how the original courses of the streams in this area were decided. The sea-level had risen to 200 m above the present level, and had drowned most of south-east England, except for an island in the Weald (Figure 5.10.) As the sea began to fall slowly towards its present height, the sea-floor was exposed as land, and rivers began to extend their courses across the relatively flat surface.

B: Early Capture of Headwaters

B. The sea-level fell still further, and the rivers began to cut deep valleys in response. Harder bands of rock, like the Chalk, caused the drainage to be concentrated into two main systems, the Blackwater in the west (B), and the Wey (W).

44

C: The Wey Becomes Dominant

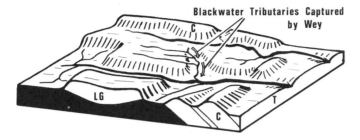

Blackwater Tributaries Captured by Wey

C. As the rivers continued to cut downwards the Wey became the more important of these two, capturing the headwaters of the Blackwater one by one, since it entered the Thames at a lower point and was therefore flowing at a lower height. The Wey also began to extend the area of its control in the easily eroded Weald Clay to the South.

D: The Situation To-day

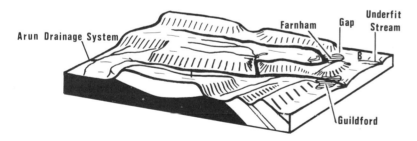

D. The present position. The Wey now controls all the headwater streams south of Farnham and Guildford, but has also lost some of its own supply to the south-flowing river Arun in the Weald Clay zone.

◄ **Figure 5.8 River capture (2).** The River Wey versus the River Blackwater. Follow the story of their history through the four diagrams A–D.

Key to the rocks of the area: L.G. Lower Greensand; C. Chalk; T. Tertiary rocks. Diagrams based on Wooldridge and Linton.

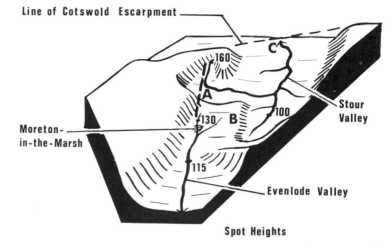

Figure 5.9 River capture (3). The river Stour versus the river Evenlode. The Stour flows into the Warwick Avon, and into tidal water within 65 km. The Evenlode, which joins the Thames and has a course of 225 km before reaching tidal water, once rose at the top of the escarpment (old course shown by line A), but now starts in a wide valley near Moreton. The Stour has cut back into the escarpment at C, has captured these old Evenlode headwaters and is likely to divert even more of its water. Can you see why? The figures on the diagram are heights in metres.

45

Superimposed Rivers

As the rivers cut downwards from the original layer of rock which determined the main features of their courses, they often cut through to underlying rocks with a more complicated structure. The drainage pattern determined by the overlying rocks is then imposed on those underneath: it is imposed from above, or superimposed. We often find this happening in Britain. The Lake District was originally a dome of rock on which a pattern of radiating streams developed (Figure 5.6), but these cut through the top layer of Carboniferous Limestone and now flow across the underlying Ordovician and Silurian rocks whose structure is quite different. But they still retain the radial pattern, which has been emphasised by recent glacial erosion.

The lowlands of Britain have a covering of glacial debris in many areas, and the irregular, tumbled nature of these deposits often led the rivers to take up new courses across their surface as the ice melted. The rivers have now cut through the soft glacial clays and sands, and are being superimposed on the solid rocks beneath.

The Weald of Kent illustrates another type of situation where superimposition has taken place, and Figure 5.10 shows you what happened there. In one way and another a good proportion of British streams have been superimposed.

The Effect of a Changing Sea-level on River Valleys

The **base-level** of river erosion is sea-level: no river can erode a valley beneath this level. As has already been indicated, however, the sea-level itself has not been very stable in the most recent period of geological time, which has seen the formation of most of our scenery. British river valleys record these changes, and a study of them helps us to illustrate the factors controlling many of the processes involved in river action.

A **rising sea-level** causes the lower parts of the river valleys to be flooded, and the rivers themselves to be slowed down. This results in the deposition of much of their load in order to build up their valley floors to profiles resembling the original. Quantities of silt are poured into the drowned part of the valley. The seaward ends of many valleys

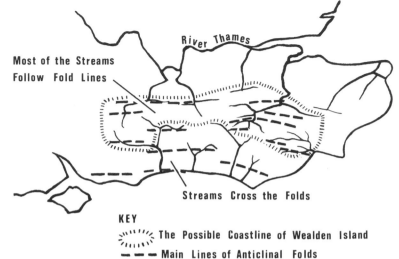

KEY

⟩‧‧‧‧‧‧⟨ The Possible Coastline of Wealden Island

— — — Main Lines of Anticlinal Folds

Figure 5.10 The superimposed Wealden rivers. At the beginning of the Pleistocene period the sea-level rose to a point 200 m above the present level, and all south-eastern England was covered except for an island in the central Weald. Within this area the rivers could flow in the original valleys, parallel to the lines of folding in the underlying rocks. The sea surrounding this island destroyed most of the drainage pattern it covered, and when it began to fall new rivers were formed, extending their courses across the exposed sea-floor. These new rivers cut their valleys across the folds and are quite unrelated to the structures of the underlying rocks.

are filled with alluvial deposits of this type. The valleys of the Sussex coast—the Arun, Adur and Cuckmere—have up to 30 m of alluvium at their mouths and flat, marshy floors for many miles inland.

A **falling sea-level** causes the rivers to cut new valleys related to the lower base-level. Starting at the mouth, and working upstream, youthful features appear, as shown in Figure 5.11(2), and the process of new erosion is known as **rejuvenation**. Even old river meanders are pre-

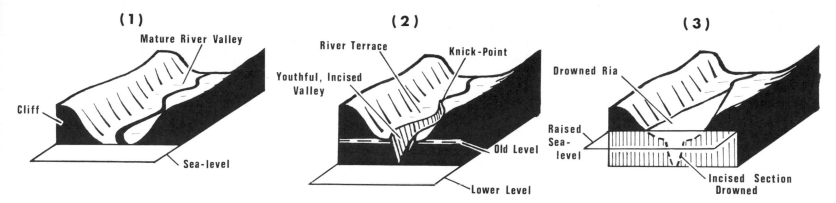

(1)
Mature River Valley
Cliff
Sea-level

(2)
River Terrace
Youthful, Incised Valley
Knick-Point
Old Level
Lower Level

(3)
Drowned Ria
Raised Sea-level
Incised Section Drowned

Figure 5.11 Changes in sea-level. When the area (1) is affected by emergence of the land (2), erosion is intensified in a narrow zone along the river's course. The old valley floor remnants are found as terraces on the valley sides, and there may be rapids, or a waterfall at the knick-point, where the new and old valleys meet. If the sea-level rises again (3), the lower part of the valley becomes drowned.

served in this renewed down-cutting as **incised meanders**. If the sea-level falls only a short distance, and the local rocks are resistant, the river will cut down more slowly and the incised meanders will be ingrown, as demonstrated in Figure 5.12; if the incision is rapid the meanders will be entrenched.

Many British river valleys show evidence of both a rising, and a falling, base-level. The Taw and Torridge valleys in north Devon, for instance, contain fine examples of ingrown meanders incised from a former valley floor which reached the sea when it was 100 m above the present level. The combined mouths of these two rivers, however, are drowned estuaries. When we consider these facts, and those mentioned above in connection with the Weald, we conclude that the following events must have taken place:

(1) In the early Pleistocene period the sea rose to 200 m above the present level.

(2) Since that time it has been falling with minor 'still-stands', such as the one at approximately 100 m. This falling sea-level caused the rivers to rejuvenate, and many were superimposed on the underlying rocks.

(3) The Ice Age caused a series of fluctuations of sea-level: it fell as the ice sheets advanced and enlarged, then rose again when they melted. Throughout this time the overall fall was continuing.

(4) At the end of the Ice Age the sea-level reached its lowest point. This must have been at least 50 m below the present level, and may have been much more. The streams were rejuvenated, and the valley mouths cut down to this level (cf Figure 5.11(2)).

(5) Finally the ice melted and the sea rose to its present position, drowning the deep gorges that had just been cut at the river mouths, and leading to deposition on the valley floors (cf Figure 5.11(3)).

Thus a picture of the recent geological history of an area can be built up by studying the landforms and drainage patterns of that area.

Antecedent Drainage

It is not only movements of the sea which interrupt the progress of

River terraces in the Findhorn Valley, Nairnshire. The river has cut down into soft morainic deposits, and the stages are recorded in a series of terrace steps on the valley sides. Draw a sketch of the photograph, numbering the terraces in order of formation, and try to calculate the height of each above the river bed. (*Crown Copyright*)

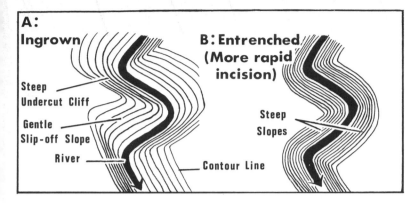

Figure 5.12 Incised meanders. The contour patterns of two types of incised meander. Why does pattern B result from more rapid erosion?

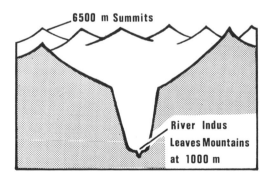

Figure 5.13 The Indus gorge. The gorge is incised 5500 m below the Himalayan summits. Steep slopes are indicative of very rapid erosion: the lower section has been cut more rapidly than the upper.

river erosion. Earth movements, volcanic action and changes of climate may cause rivers to change course, or to cease activity altogether: many of the deserts once had rainy climates, when rivers carved valleys which can still be seen.

Sometimes a river may not be diverted by earth movements, and may continue its course across an uplifting group of mountain ranges. It will have to erode rapidly in order to keep pace with the uplift, and there may be times when it is temporarily ponded back to form a lake until that overflows down the old valley. It is thought that the Colorado river in the western USA is of this type, and that the rivers which cross the Himalayas into north India have continued to cut deep trenches as the mountains have risen across their path. The Indus, for instance, has a gorge 5500 m deep (Figure 5.13), and the lowest part is very steep-sided as the rate of uplift has increased in more recent times. These rivers came before, or antedated, the mountains, and are known as antecedent. They emphasise the slow rate at which mountain uplift proceeds.

River Transport

A river may be regarded as a 'conveyor belt' along the bottom of its valley. All the debris moved down the slopes by soil creep and by landslides falls into the river and is carried down the valley towards the sea. This **'load'** consists of a large soluble fraction; the more visible silt and sand carried in suspension; and the pebbles on the river bed, which are moved in time of flood. When the river moves faster it can carry more debris. If a river's speed is doubled it will be able to carry sixty times as much debris.

River Deposits

A river flowing across a lowland area has a more gentle gradient and is slowed down. The waters spread out in a wider river bed, and friction is increased, causing the load to be dropped.

The extreme case of this effect is seen in arid areas after a storm. It was not very long ago that these areas had a much higher rainfall, and

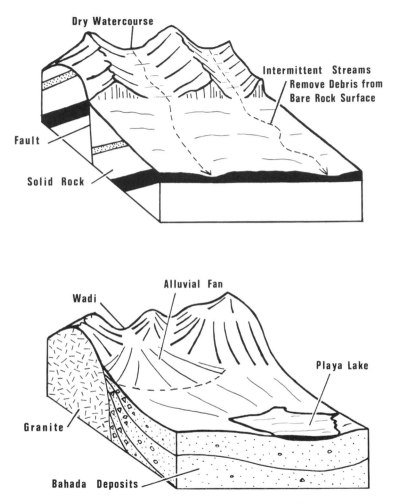

Dry Watercourse

Intermittent Streams Remove Debris from Bare Rock Surface

Fault

Solid Rock

Alluvial Fan

Wadi

Playa Lake

Granite

Bahada Deposits

Figure 5.14 Pediment and bahada. Transport and deposition of debris in a dry region. Can you explain the differences between the two types of terrain?

the valleys cut by the rivers at that time provide channels for the great sheets of water rushing down from mountainous areas. The streams collect vast quantities of loose rock fragments from numerous small gullies and normally dry valleys. On reaching a plain at the mountain foot the water spreads out, rapidly dropping all the debris, and forming a fan-like structure. When this occurs in an area of inland drainage, the lowlands will be filled in with these **bahada** deposits, and the water may drain into a **playa lake**, which is dry for most of the time since the water is soon evaporated. If the river is able to drain into the sea it will carry away the debris across a bare-rock **pediment** surface. Both of these possibilities are shown in Figure 5.14. Similar rapid accumulation of river deposits has taken place at the foot of newly raised mountain ranges. When the Alps were first formed there were deep depressions to north and south, but the rapid erosion of their high peaks filled in the areas we now know as the central lowlands of Switzerland and the Po valley of northern Italy.

Most river deposits are associated with the normal narrow channel, and are commonly found near the mouths of rivers, as illustrated in Figure 5.15. The valley deposits are very varied because of the changes that are always taking place in the flow and load of a river, and in its course over the flood-plain. The deposits of the river bed are usually coarse, lens-shaped bodies of gravel mixed with sand and clay. These are the materials which the river moves only in time of flood, and are known as a lag deposit. When a river overflows its channel it builds up levée banks since the coarser load is dropped as soon as the current slackens. On the other hand the deposits resulting from the river spreading across the whole flood-plain are fine silts and muds. Finer deposits also occur in cut-off (oxbow lake) sections of the river; the still waters soon fill with sediment and organic matter as plants begin to colonise the pool. Terraces along the valley side are old valley-floor alluvial deposits, left by rejuvenation, and are often covered by a thin veneer of gravel and well-drained soils.

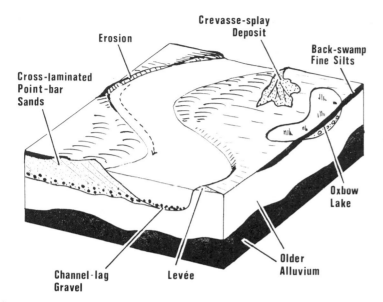

Figure 5.15 Deposits left by a river meandering across a flood-plain. Notice the sequence of deposits formed on the inside of a meander and in a cut-off lake.

Deltas and Estuaries

Deltas and estuaries occur at river mouths. Deltas form where a vast quantity of sediment is brought down by the river, and estuaries are found where the river sediment load is small.

Deltas. Deltas form as a river's load is checked on entering the sea or a lake. The load is jettisoned, and, if the supply of debris is too great for

Milford Sound, South Island, New Zealand. Notice the pattern of distributaries, the underwater pattern of deposition, and the growth of vegetation. This delta is at the mouth of a typical glaciated valley in the Southern Alps. (*White Aviation Ltd*)

51

the waves or offshore currents to remove it, a delta will build up and extend the land seawards. Many deltaic features are determined by the relationship of the river water to that of the sea or lake into which it flows:

(1) If it enters a freshwater lake like the Caspian Sea, the river water fans out, dropping its load rapidly in a broad arc. Until recently the river Volga delta was advancing 1 km into the Caspian every three years, but has now entered deeper water.

(2) If, however, the river is carrying a great quantity of load, it will form a dense, mud-charged current, which will dive beneath the clear surface waters. An example of this is the River Rhône, which is heavily laden with rock debris washed out of Alpine glaciers as it enters lake Geneva: it has carved a deep trench in the lake floor, leaving only a small delta. The sediment accumulates on the deep lake floor, and the river is clear as it leaves the lake.

(3) When river water enters the sea it is usually less dense than the seawater and flows over the surface, spreading outwards. The water at the margin of each channel is slowed, and coarse levée banks are built above the flood-plain and sea-levels. The Mississippi delta, seen in Figures 5.16 and 5.17, illustrates this effect with the 'bird's-foot' shape of its present delta. Its deposits are thousands of metres thick and have been likened to a pile of leaves, each of which reflects a separate course of the river across its delta and the deposits associated with these courses: the veins of the leaf represent the coarse channel and levée deposits and the leaf flesh the fine swamp deposits between.

As a channel is built outwards the gradient becomes lower until the river cuts through (or crevasses) its banks higher up and takes another course, probably cutting across old swamps and building out over marine sediments. Seven deltas of the Mississippi have been recognised for the last 3000 years, but its course is now artificially regulated. The weight of the sediments—the river delivers 1 km³ every three years—causes subsidence and compaction followed by the drowning of abandoned sections of the delta.

Look at maps of other deltas and see how the river deposits, sea action and subsidence have moulded the shape.

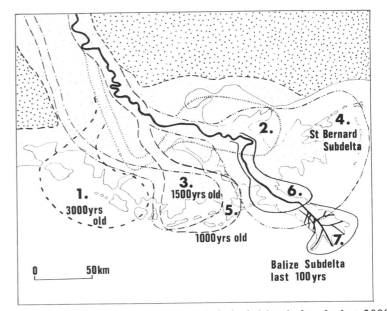

Figure 5.16 The development of the Mississippi delta during the last 3000 years. What has happened to the older delta areas? Make your map of the types of deposit you might expect to find over the area (ie fine mud, silt, sand).

Estuaries. Estuaries occur at the mouths of rivers bringing restricted quantities of sediment into the sea. Many river mouths have wide openings due to postglacial drowning. The Thames estuary is one of the most-studied examples, but most British river mouths have this character.

Figure 5.18 shows the contrast between the sediments of the outer estuary of the Thames and those of the inner estuary. The inner estuary muds are deposited where the freshwater meets the salt water: the river flow is checked between Waterloo Bridge and Woolwich, and individual clay particles in the freshwater are able to flocculate (ie cling together) and sink to the river bed. Here there is a deposit of

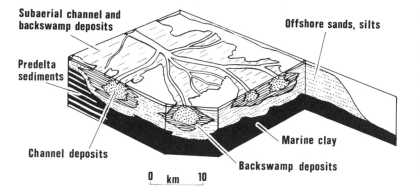

Figure 5.17 The features and deposits of a typical delta. Describe the sequence of deposits which would be found if a well was drilled through these deposits in a position just offshore, and compare the sequence found in a river flood-plain (Figure 5.15). Vertical lines on diagram = 200 m.

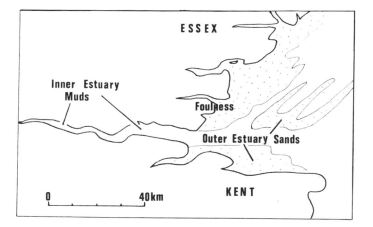

Figure 5.18 The Thames estuary: its shape and deposits.

fluid, slimy mud, which may be dispersed in the next low tide by strong channel currents, but which builds up very rapidly in sheltered hollows. Dredged tanker channels have to be continuously emptied, the greatest effect being in winter, when the river brings down more mud in suspension.

In the outer estuary the sands cover an old topography of channels cut deeply towards the low sea-level of the late glacial phase. The sands have spread in from the North Sea glacial deposits and from the deposits of the Rhine, and are arranged in banks separated by tidal scour channels. The deposits shift, filling the tidal channels, which are floored with coarser lag deposits (pebbles, shells, wood, old boots).

The study of these estuarine deposits is important in connection with such projects as the flood barrage and the reclamation of the Foulness area. It is important to know, for instance, whether the flood barrage will shift the zone of mud formation downstream and cause trouble for the new port developments. The reclamation of Foulness will be made difficult and costly by the variable nature of the underlying, water-laden sediments (Figure 5.19). It would also restrict water flow in the estuary, which could also affect the dock areas.

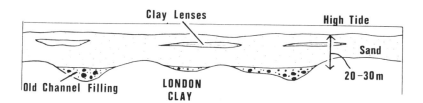

Figure 5.19 A section through the Thames estuary deposits at Foulness. The length of the section is 1 km. If water was extracted from the sands compaction would take place. Suggest how this might affect the surface of the deposits.

6 Lakes and Oceans

great 'settling tanks', in which rivers drop their loads of sediment to emerge as clear streams carrying only the finest matter.

Lakes are essentially water-filled hollows, and the chart, Figure 6.1, gives a list of the different types of origin of these hollows together with a series of examples of the more important types.

LAKES

Lakes are features of a youthful river system, and are doomed to be filled in with rock debris and to disappear. They can be regarded as

Lacustrine Deposits

The main process taking place in a lake is deposition. It is common to find that deltas form where rivers enter a lake, and some have cut lakes in two by extending their deposits right across, as in the dividing

THE ORIGIN OF LAKE BASINS

Origin	Examples	Origin	Examples
EARTH MOVEMENTS Crustal warping Backtilting of valley Rift valley Tear fault	Caspian Sea; Lake Victoria; Lough Neagh Lake Kyoga (upper Nile) Lake Tanzania; Great Basin (USA) Lac de Joux (Jura Mountains)	DEPOSITION Morainic barrier Irregularities in glacial drift ('kettle holes') Ice barrier Landslides, mudflows River deposits—ox-bow lakes Coastal lagoons enclosed by bars, sand dunes	Lake Garda (northern Italy) Glaciated plains, valleys—eg Tweed Marjelen See (Alps); Glen Roy (Scotland) Lower Mississippi valley SW France; Chesil Beach, Dorset
VULCANISM Rings of volcanic ash Crater, caldera Lava barrier across valley	Maare of Eifel Mountains (Germany) Crater Lake, Oregon (USA) Sea of Galilee		
EROSION Glacially eroded hollows Ribbon lakes Cirque-floor tarns Wind deflation Solution hollows	Canadian Shield; Finland Thirlmere, Lake District Bleawater, Lake District	ORGANIC Coral reef barrier Beaver dams Man-made dams Man's excavations—gravel pits, peat-digging	Pacific Ocean Yellowstone National Park (USA) Kariba (Rhodesia) Norfolk Broads
		METEORITE IMPACT	Ashanti Crater (Ghana)

NB Many lakes are caused by a combination of factors, eg Loch Ness is a glacial hollow along a fault-line; the Sea of Galilee is in a rift valley but is dammed by a lava-flow; the Great Lakes of North America were hollowed out by ice tongues, dammed by the terminal moraines and uplifted at the northern end by isostatic compensation.

Figure 6.1

stretch of land between Buttermere and Crummock Water in the Lake District. The coarsest fragments are dropped near the river mouth, but sand and mud may be carried far out into the lake. Organic matter, particularly shells, but also much plant debris, becomes more important in the centres of large lakes. Shallow lakes in cold areas, such as northern Sweden, often have a rather frothy deposit of 'bog iron ore' around their edges.

Varved clays are the result of seasonal deposition in a lake supplied with glacial meltwater. The deposits on the floor of Lake Zurich in Switzerland, which are carried into the lake by rivers descending from the Alpine glaciers, have an alternation of thin, dark layers of mud rich in organic matter, and thicker, lighter-coloured sands containing calcium carbonate. In winter the rivers supply very little glacial debris, but in summer the light-coloured layer is formed of a very high proportion of 'rock-flour' brought from melting icefields. The best-known varves are those formed in lakes trapped between melting ice masses and ranges of hills. Their formation is illustrated in Figure 6.2.

It has been possible to use varves to date the retreat of the last ice-sheet across northern Europe. They differ in thickness from year to year and distinctive patterns are thus built up which can be traced from the site of one glacial lake to another.

Salt lakes are important areas of chemical deposition in dry areas, especially where drainage is to an inland basin, and not to the sea. As the water is evaporated from the lake the salts contained in the water become more and more concentrated, and eventually the remaining solution will become saturated with a particular salt which will be precipitated. The Dead Sea (really an inland lake) has nearly ten times the concentration of salt in seawater. The type of salt deposited in the inland lakes varies according to what has been supplied to the lake from the surrounding rocks and also to the intensity of evaporation. The Great Salt Lake of Utah is surrounded by layers of carbonate salts, formed when the lake covered a greater area, and sodium chloride is being precipitated on the floor of the present lake. Farther south, in Nevada, there are a series of small playa lakes in which

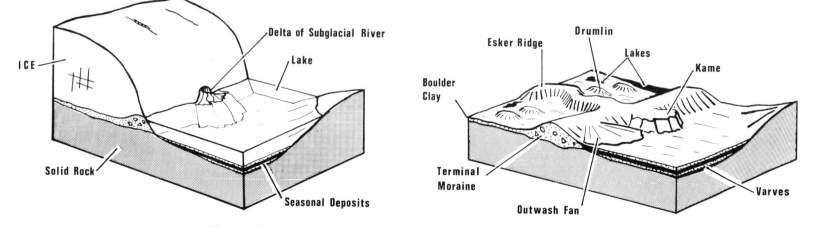

Figure 6.2 Ice sheet deposits. The diagrams show the formation of distinctive deposits beneath the ice, and in meltwater lakes in front of the ice sheet.

sodium carbonates are being deposited. One of the most interesting of these inland lakes, the Gulf of Karabugaz by the Caspian Sea, is illustrated in Figure 6.3.

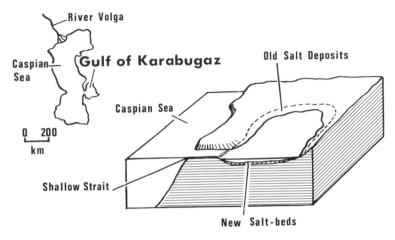

Figure 6.3 A salt lake. There is a constant flow of water through the narrow strait from the Caspian Sea. The water, only one-third as salty as ocean water, is evaporated in the shallow gulf and becomes nearly ten times as salty as seawater: this leads to the precipitation and deposition of layers of salt on the floor of the gulf. Calcium carbonate and gypsum are overlain by sodium sulphate.

Seawater may be trapped in the coastal lagoon of an arid area, and when this happens there is a consistent pattern of precipitation, since seawater has a relatively constant composition, including 3·5 per cent of dissolved matter, mostly sodium chloride.

(1) When half the seawater has been evaporated, the iron oxides and calcium carbonate are precipitated;

(2) When four-fifths of the water has been evaporated, gypsum is precipitated;

(3) When nine-tenths of the water has been evaporated, sodium chloride is precipitated;

(4) Finally, if evaporation continues, the bitterns (magnesium and potassium salts) are deposited.

This order faithfully reflects the solubility of these minerals. Sea salts have been deposited in this way in the Piano del Sale on the border between Abyssinia and Eritrea. This was once a gulf of the Red Sea, but a series of lava flows cut it off, and it has become a great evaporating dish in the desert climate. It is lined with gypsum and rock salt covers the centre of the basin, an area 30 km across. In the very centre, on top of the rock salt (sodium chloride) are layers of potash. The original basin was 300 m deep, but that depth of seawater would only have left a 5 m layer of salt. There must have been a series of invasions by the sea before the pile of lava flows at the entrance reached its present height. In fact repeated invasions by the sea or a continuous influx of seawater insufficient to balance evaporation have been required to produce the great accumulations of these salts such as the Stassfurt deposits in Germany (Figure 8.13) or the recently discovered ones in north-east Yorkshire (Figure 17.8).

SEAS AND OCEANS

Seventy-one per cent of our globe is covered by water. Eighty per cent of the oceanic area is over 3000 m deep. This vast realm is one of the most difficult for men to study, and our knowledge of it is increasing very slowly. Many nations, however, are now making intensive efforts to investigate what lies beneath the ocean waters, and are using many new techniques. Their findings will be of great interest to geologists, since not only is this a large part of the planet we are studying, but nearly all the sedimentary rocks are formed beneath the waves and most of the fossils entombed in the rocks lived in the sea.

The Features of the Ocean Floor

Shallow seas—often referred to as **epicontinental** seas—are found covering the margins of our continents, as shown in Figure 6.4. This is where the deposition takes place which will eventually give rise to new rocks and land. Such continental-shelf areas surround most continents,

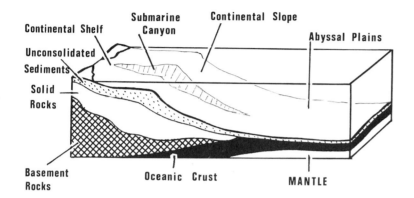

Figure 6.4 Zones of marine deposition. The main sections of the sea-floor, showing some of the relief features near the continental margin, and the main zones of deposition. The littoral zone is uncovered at low tide; the neritic zone is on the continental shelf; the bathyal zone is on the continental slope; and the abyssal zone is on the ocean floor.

but vary in width from several hundred km to nothing. The **continental slope** begins at depths of 150–200 m below the sea-level, and has a gradient of 1 in 100 near the top, flattening to 1 in 1000 as it joins the abyssal plains. **Abyssal plains** are flat areas covering a large proportion of the ocean floor, and they are connected through gaps in the submarine mountain ranges. Their extreme flatness may be due to the spreading out of debris after it has slumped down the continental slope.

The edges of the continental shelves are often deeply cut by **submarine canyons**, which extend down across the continental slope.

These canyons may be incised as much as 1000 m into solid rock and sediment, and have steep walls and a branching system of tributaries. Many follow the lines of river systems out to sea (eg the River Hudson which enters the sea at New York and the River Congo in Africa), but some have no such connection. It has been suggested that these canyons were carved by rivers extending their courses across the continental shelf to a low, Ice Age, sea-level. Some, however, extend down to 3000 m below the present level, and the sea could not have fallen so far. Recent investigations have shown that seawater charged with suspended sediment stirred up by a storm often rushes down the continental slope, causing damage to cables, and even ripping through rocks. These currents of water are known as **turbidity currents** and may be the main factor at work in the production of submarine canyons.

The deeper parts of the oceans have a considerable relief with vast, flat plains broken by troughs and peaks. **Oceanic rises** may cover hundreds of square km, and occasionally break the surface. The Rockall Rise in the North Atlantic is one such feature. **Seamounts** (known as guyots in the Pacific Ocean) are isolated peaks now covered by up to 1000 m of water. They are possibly volcanic in origin, and may have had their top-most points cut off by erosion, or the weight of the volcanic rock may have caused the underlying crust to subside. Volcanoes are common in oceanic areas; they have to erupt a vast quantity of material in order to raise their heads above sea-level and then maintain them there. The lack of certainty on the origin of these features reflects the poor state of our knowledge concerning many submarine relief forms. Ocean surveys have increased greatly in intensity since 1960, and their results are becoming known.

The massive **Mid-Atlantic ridge**, occupying nearly one-third of the ocean width, was thought to be unique, but recent surveys have located world-wide continuations extending into the Indian and Pacific Oceans, although neither of these are centrally placed in their oceans (Figure 6.5). The features of the Mid-Atlantic ridge include a double central range separated by a rift valley and paralleled on either side by a series of other parallel ranges getting gradually lower as one travels

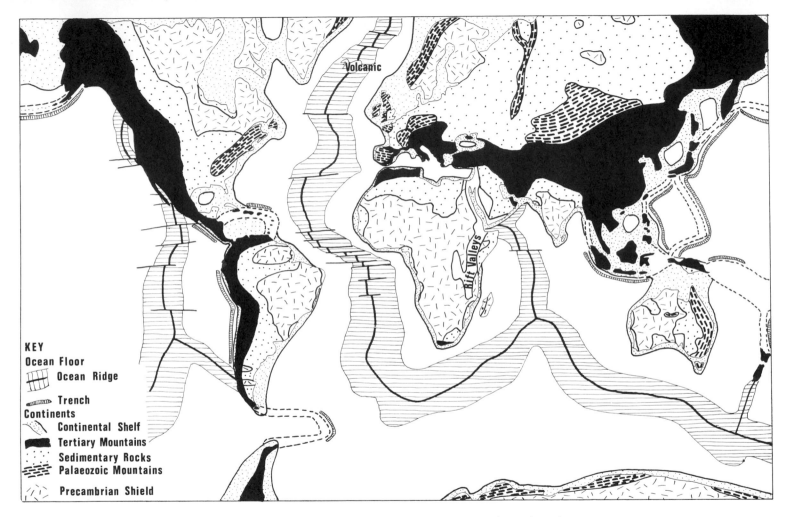

KEY
Ocean Floor
Ocean Ridge
Trench
Continents
Continental Shelf
Tertiary Mountains
Sedimentary Rocks
Palaeozoic Mountains
Precambrian Shield

Volcanic

Rift Valleys

Figure 6.5 The major structural regions of the ocean basins and continents.

away from the central ridge. Wherever peaks extend above the ocean surface they are associated with volcanic activity, and the whole ridge is probably formed of volcanic rock. Earthquakes are also associated with the zone (map, Figure 10.1), which is obviously an active part of the Earth's crust. The East Pacific ridge is much less marked in relief, but has similar parallel ridge features extending far out across the ocean-floor on either side. It even passes beneath the Californian coast.

Oceanic **trenches** are the deepest parts of the oceans, and often extend to twice the depth of the average ocean-floor (the deepest trench is over 11 000 m below sea-level, Figure 2.4). The trenches are thousands of kilometres long, but only 100–200 km across, and are found lying parallel to chains of volcanic islands (eg south of Java–Sumatra; east of Philippines–Japan; south of Aleutians), or ranges of mountains (eg west of South American Andes). They are also associated with a zone of earthquake activity: this reflects the deep-seated release of stresses in the Earth's upper layers. (Earthquakes are studied in Chapter 13.)

Seawater
We all know seawater is salty. The chemicals come from the minerals carried in solution by the rivers. Almost all of the calcium carbonate entering the sea is removed from solution by marine organisms. Seawater is thus a solution of the very soluble salts which are not used by sea creatures. Sodium chloride is the main one (over 77 per cent), followed by magnesium chloride (11 per cent), and small proportions of other magnesium, calcium and potassium salts.

The proportions of the various minerals change little throughout the world, but the actual concentration of the salts (the **salinity**) alters from one sea to another. This is because of the great differences in amounts of rainfall, rates of evaporation, supplies of fresh water and communications with the open oceans. Most of these oceans have salinity of 35 parts of salt per thousand; the Red Sea has 38·8; the Baltic Sea has 7·8. Can you account for these last two figures?

Seawater also contains, especially near the surface, a certain amount of oxygen, which is vital for marine animals. Plants can live only where sunlight can penetrate and allow photosynthesis to take place. The proportion of carbon dioxide increases deeper in the oceans, and in stagnant basins such as the Black Sea and some deep fiords there is no oxygen at the bottom. The Black Sea conditions are illustrated in Figure 6.6.

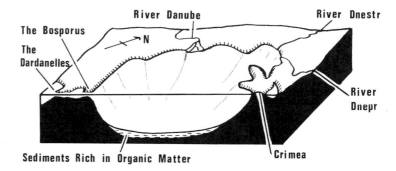

Figure 6.6 The Black Sea. There is a shallow outlet via the Bosporus, but the deepest parts are 2300 m below the surface. Whilst the surface layers are rich in life the stagnant waters beneath are without oxygen, and undecomposed organic matter accumulates on the sea-floor.

Ocean-water Movements
Large-scale movements of water masses in the oceans are known as **currents**, and are caused by differences in the temperature and salinity of the water. Cold water is more dense than warm, and drifts towards the Equator from the Poles, sinking beneath the surface layers of lighter tropical water which is moving towards the Poles. Figure 6.7 shows the circulation in the Atlantic Ocean. **Drifts** are surface movements of the water caused by the prevailing winds, and, like the currents, are modified in their courses by the shapes of the continents. The North Atlantic Drift, for instance, is closely connected with the westerly winds of that zone, and there is a two-way effect: whilst the winds cause the water to move, the relatively warm waters help to make the air passing over Britain milder than it would otherwise be.

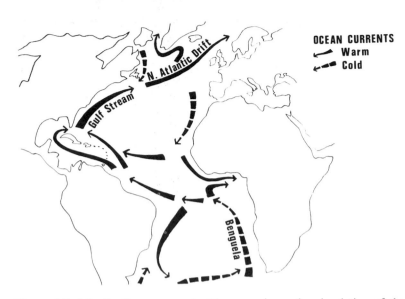

Figure 6.7 Atlantic Ocean currents. The map shows the circulation of the surface waters. There are also deep currents of cold, dense water moving from the polar regions: these are sometimes drawn to the surface as the waters move away from one landmass towards another, and form cool currents on the west sides of continents.

Tides are oscillations of ocean water caused by the attraction of the Moon and Sun. They are scarcely noticed in the open oceans, but when the movement of water is constricted in a narrow channel strong local currents may be experienced, as well as a great difference between high- and low-tide levels. The tidal range at Liverpool is 10 m, and the strong tidal currents help to keep the Narrows at the port entrance clear of silt. Some of the world's greatest tidal ranges are found in the Bay of Fundy. Look it up on an atlas map, and explain why this is so.

Waves are caused by the friction of the wind blowing across the surface of a body of water. They may grow to 16 m in height, but anything over 10 m is unusual, even in a strong gale. The wave height will depend upon the strength of the wind, the depth of water, and on the fetch—the distance of water over which the wind has travelled. The largest waves will occur in deep water far from land during a gale. Each wave is just an upheaval of water, and seldom involves actual movement of the water in the direction towards which the wind is blowing. The particles of water move in a circular path as the wave passes through them, but this effect dies out with depth until there is just a gentle to-and-fro movement. Wave base is up to 200 m beneath the surface.

As a wave enters shallow water it is slowed down, the wave length becomes shorter, and its height greater. This makes the mass of water unstable, and eventually the top of the wave collapses in a breaker. Figure 6.8 shows that there are two main types of wave, which either help to build up a beach, or will carry much of the beach material out to sea. If the waves meet the shore at an angle they will cause debris to be moved along the coast in the manner illustrated in Figure 6.9.

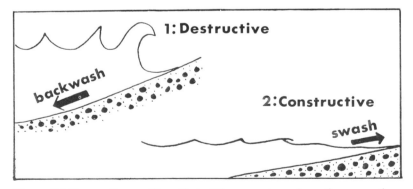

Figure 6.8 Types of wave. The effect of the wave depends on the proportions of swash, which sends material up the beach, and of backwash or undertow, which drags it out to sea.

Many holiday resorts make use of this process of **longshore drift**, and build groynes, or breakwaters, to retain some of the moving pebbles. These protect the promenade, but lead to increased erosion farther along the coast, from which the protective pile of pebbles is removed.

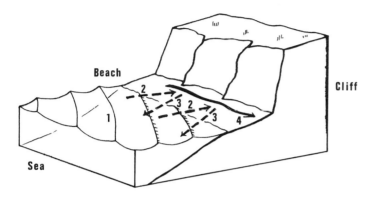

Figure 6.9 Longshore drift. Longshore drift causes pebbles to be moved along the beach as follows. 1—the waves meet the shore at an angle; 2—pebbles are carried up the beach in the direction of wave movement; 3—they roll back down the steepest slope taken by the undertow; this happens many times and the result—4—is that there is an overall movement of rock debris along the beach.

Marine Erosion

The most obvious and dramatic result of wave action is found in the wearing away of coastal rocks to form cliffs. The coast of Yorkshire between Flamborough Head and the Humber estuary has been pushed back 9 km in the last 1800 years as the sea has washed away the soft cliffs of unconsolidated glacial deposits: over thirty villages have disappeared. Even cliffs of resistant rocks may be worn back at such a rate that short rivers emptying into the sea may be continually rejuvenated. Coasts are eroded most rapidly when they face into the full force of the waves. One of the best examples of cliff-retreat is the coast

of north-west Devon and north Cornwall between Hartland Point and Bude, a section of which is sketched in Figure 6.10. The brown sandstone cliffs north of Sandown, on the Isle of Wight, also demonstrate the speed of cliff erosion, for a brick field-boundary wall was overhanging the cliff top by 2 m until the 1964–5 winter, when it eventually crashed down to the beach beneath: the sea had gradually worn away the cliff-base until the wall collapsed.

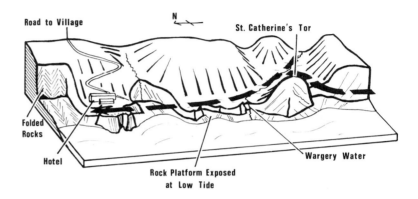

Figure 6.10 Hartland Quay, North Devon. Marine erosion is rapid along this coast, which faces the full force of the Atlantic rollers. The old valley, marked by the broken arrow, has been cut into and dismembered, and streams are often left hanging above the beach in waterfalls.

Solution affects some cliffs, especially those formed of limestone, and much of the chalk around Flamborough Head is pitted with solution hollows. Most erosion is accomplished by the sheer force of the breakers, compressing air in rock fissures, and hurling rock fragments against the cliff base. The force of waves in winter storms is great enough to shift huge blocks of rock, and the general effect of such wave action is to undermine the cliffs by a horizontal sawing mechanism. The rock fragments involved are soon worn down to rounded pebbles or cobbles.

Beachy Head, Sussex. What features of this coast show that the chalk rock is being eroded by the sea, and the cliffs are retreating? (*Aerofilms*)

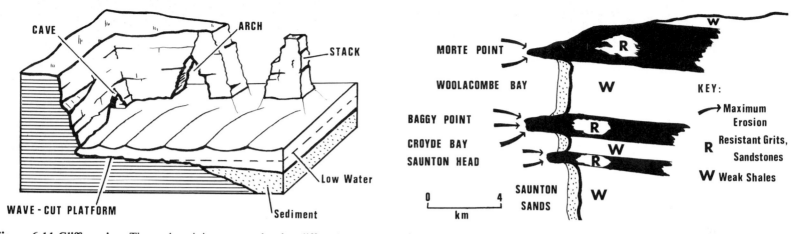

Figure 6.11 Cliff erosion. The undermining sea erodes the cliff rocks at sea-level, opening out caves and arches along lines of weakness.

Figure 6.12 Differential marine erosion. The headlands, subject to the greatest wave attack, will eventually be cut back to form a straight coastline.

As marine erosion proceeds, and the cliff-line is pushed back, a **wave-cut platform** appears at low tide. This suggests that most of the erosional undercutting takes place at a point between high and low tide, and many cliffs demonstrate the zone of maximum erosion in a distinctive notch, as shown in Figure 6.11. Eventually, when the waves have to travel for some distance in the shallow water above the rock platform, the process of wearing back the cliff-line slows down. The 'cutting edge' of the sea becomes less effective, and the cliff face may be moulded to a greater extent by rainwater gullies and general weathering.

The details of your local cliffs depend on a number of other factors as well, including the relief of the area, and the type of rock being attacked. If the area is low-lying the cliffs will be far less imposing than if it is hilly upland. When an alternating series of hard and soft rocks is

tilted into a vertical position, the sea will wear away the outcrops of rock at varying rates and will carve bays in the weak rocks, leaving the more resistant as headlands. This has happened where the western end of Exmoor in north Devon meets the sea, as shown on the map, Figure 6.12.

The sea supplements the undercutting of the cliffs by working on the weakest places in the rocks, such as joints, fault-lines and bedding planes (joints and faults are described in Chapter 12; bedding planes in Chapter 8). Any gap in the rock is liable to wave attack. **Caves** are gouged out (Figure 6.11), and two formed back-to-back along the same line of weakness may combine in an **arch**, which eventually collapses to leave a **stack**. When these isolated rock masses are removed the sea will have pushed back a portion of the coast. Den's Door, just north of Broadhaven, Pembrokeshire, has resulted from this

Stacks and 180 m cliffs at Duncansby Head, Caithness. The rocks are Old Red Sandstone. (*Crown Copyright*)

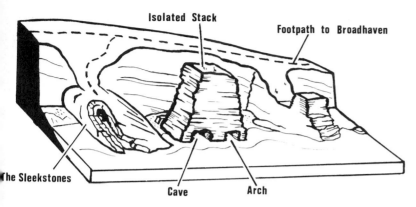

Figure 6.13 Den's Door, Pembrokeshire. This feature, just north of Broadhaven, illustrates the undermining effect of the sea. The Sleekstones is an overfold in the Coal Measure rocks.

series of events: it has been isolated as a stack, and now the sea is undermining it (Figure 6.13).

Thus the erosional force of the sea is concentrated at sea-level, between tides, and any wearing away carried out at greater depths is insignificant. Swift currents rushing through narrow straits, as between Dover and Calais, merely prevent deposition taking place, and do not carry out erosion.

Sea and Ocean Deposits

The material supplied to the seas by rivers and coastal erosion is transported in solution, in suspension and by rolling along the sea bed. Smaller quantities come from glaciers, or are blown over the sea. It is to be expected that the larger fragments will remain close to the shore, whilst only the mud fraction will find its way far out to sea. The waters of the Atlantic Ocean are coloured by River Amazon mud 320 km

offshore, but it is exceptional for the amount to be so great that it becomes so visible.

1 Shallow Water Deposits. These are laid down between the shore and the edge of the continental shelf. Those formed between the tides are known as **littoral**; other shelf sea deposits as **neritic**. Figure 6.14 is a general summary of the nature of these deposits. Much of their

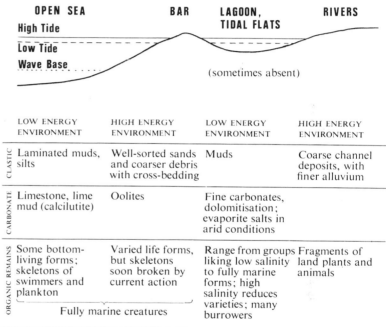

	OPEN SEA LOW ENERGY ENVIRONMENT	(BAR) HIGH ENERGY ENVIRONMENT	LAGOON, TIDAL FLATS LOW ENERGY ENVIRONMENT	RIVERS HIGH ENERGY ENVIRONMENT
CLASTIC	Laminated muds, silts	Well-sorted sands and coarser debris with cross-bedding	Muds	Coarse channel deposits, with finer alluvium
CARBONATE	Limestone, lime mud (calcilutite)	Oolites	Fine carbonates, dolomitisation; evaporite salts in arid conditions	
ORGANIC REMAINS	Some bottom-living forms; skeletons of swimmers and plankton	Varied life forms, but skeletons soon broken by current action	Range from groups liking low salinity to fully marine forms; high salinity reduces varieties; many burrowers	Fragments of land plants and animals
	Fully marine creatures			

Figure 6.14 Shelf sea sediments. Relate the position of each zone to the energy available (ie degree of wave and current action) and to the sediments produced.

nature depends on the supply of rock debris eroded from the land: if it is adequate clastic sediments will prevail, but as it declines the carbonate element in particular will predominate. The two opposite ends of the scale (ie all clastic or all carbonate) are summarised in the chart, but there are endless variations between. The lagoon and tidal flat section may be absent, so that there is no break between the bar deposits and the coast: these then become the beach.

Beaches are generally formed as banks of sand or pebbles between tides, as illustrated in Figure 6.15. Debris from cliff erosion is drifted along the coast and built up by the waves into ridges of rounded pebbles or flatter stretches of sand and mud. The constituents are largely controlled by the types of rock found locally, and the steepness of the beach is determined by the size of the debris contained in it.

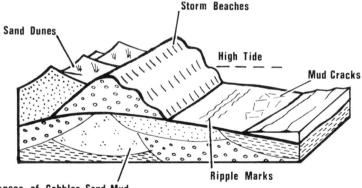

Figure 6.15 Beach deposits. Notice the type of material included in these deposits, and the size and extent of the structures affecting them. Make your own observations of such features when you visit the coast.

Where the shore slopes at such an angle that the sea is shallow, the waves are forced to break out to sea, and a **shingle bar** will be built up, enclosing a **lagoon**. **Spits** are produced on headlands when pebbles drifted along the coast accumulate in a region of slack water at a turn in the shoreline. Hurst Castle Spit, at the western entrance to the Solent (Figure 6.16), has been formed in this way. Such features result from a tendency of the sea to even up the coastline: bays are cut off and filled in, and estuaries are constricted by spits.

Figure 6.16 Hurst Castle spit. Longshore drift from the west has supplied the pebbles, and the spit has extended out into the Solent estuary as far as the incoming waves will allow. Storms often carry away the end of the spit, but it is always rebuilt in quieter periods. Marshes are filling in the shallow waters behind the spit.

When a lagoon is present it may give rise to a wide range of sediments: the common factor is their accumulation in a zone protected from wave action. The largely undisturbed and shallow waters are warmer than the local seas. They are colonised by many animals and plants, and soon fill with sediment unless scoured regularly by tidal channels. Some turn into swamps and vegetation may take over to convert them into boggy, and later peaty, areas (eg the lagoon of Slapton Ley in south Devon). In arid regions, where little land-derived

Hurst Castle Spit. Compare this photograph with Figure 6.16. This spit has been built of debris eroded from the distant cliffs near Bournemouth. (*Aerofilms*)

sediment reaches such lagoons, they become the site of chemical precipitation and high salinity in contrast to the brackish (ie seawater diluted by freshwater) state of lagoons in humid regions. Continued evaporation will lead to the formation of evaporite salts in the manner described earlier in this chapter. Such arid lagoons are known as **sabkhas**.

The sediments of the **open sea-floor** are also varied, but divisions can be distinguished in the sectors above or below wave-base. Apart from the immediate vicinities of river mouths, where estuarine muds may accumulate, the current action and water turbulence associated with wave movements, keep sorting the bottom sediments. Fine silts

and muds do not have a chance to settle and sands are generally the dominant sediments: they often contain glauconite, a green silicate of iron which forms in the chemical environment existing at this level in the sea. In carbonate seas oolites are formed, and these are common on the floor of the shallow south-western margins of the Persian Gulf at the present day, coating sand-grains from the nearby desert and tiny shell fragments with calcium carbonate. As these are rolled to and fro on the shallow sea-floor they assume a rounded shape, and are known as ooliths (Greek 'oon'-egg; 'lithos'-stone). On the adjacent coast, formed largely of recent limestones which have been raised above sea-level, the porous rock is invaded by the sea between tide levels, and much calcium carbonate is dissolved and replaced by magnesium salts to form dolomite (a mixture of magnesium and calcium carbonates).

The continental shelf off the eastern coast of the USA has a covering of recent sediments, and is the product of successive layers of rock built on top of each other (Figure 6.4). The more northerly parts have a coating of gravels, sands and muds, mostly derived from glacial debris and resorted by the sea as the level rose at the end of the last glacial advance. The central part is dominated by sandy sediments, and there is a notable contrast in the south off the coast of Florida, where land-derived sediment is scarce and the warmer waters are saturated with calcium carbonate. Shell-banks, calcareous sands and muds, and coral reefs are common.

Coral reefs are particularly interesting features from a geological point of view. Not only do they help us to understand the conditions in the past, and the method of formation of an important group of rocks, but the problem of their origin illustrates geological ways of thinking. An argument has been in progress for several years, but it now seems the problem has been almost solved.

Coral reefs are built by lime-secreting organisms with very special demands as to their living conditions. They need sunlight and warmth, with a temperature of at least $18°C$, though they are at their best between $23°$ and $25°C$. Most rapid growth occurs in the agitated, oxygen-rich surface waters. Some reef corals have been found growing

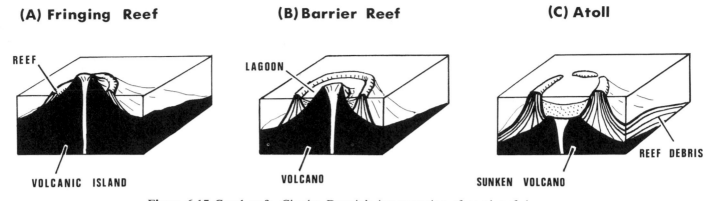

(A) Fringing Reef   **(B) Barrier Reef**   **(C) Atoll**

REEF

LAGOON

REEF DEBRIS

VOLCANIC ISLAND

VOLCANO

SUNKEN VOLCANO

Figure 6.17 Coral reefs. Charles Darwin's interpretation of coral reef development. Assuming that the reef would begin to develop near the volcanic island, he concluded that the island must have subsided for the reef to become separated from it—B—and for atolls—C—to be formed. As the island slowly sank, the reefs would be built up to sea-level, or would cease to exist.

as deep as 60 m below the surface, but they are isolated, poor specimens. Muddy water also inhibits growth. Perhaps the strangest feature is that corals grow most strongly in the zone of wave action just above low tide (ie the zone of greatest marine erosion), and are most easily worn away in quiet waters.

The largest reefs are in the Pacific and Indian Oceans. This is where Charles Darwin studied them, and where he suggested that the Fringing Reefs, Barrier Reefs and Atolls, shown in Figure 6.17, were three stages in a common series of events. Other scientists pointed out that the valleys on volcanic islands still remaining above sea-level confirmed this subsidence theory because they were drowned at their mouths. Darwin's theory did not take into account the recent sea-level changes which have been so important in forming our relief. We have already seen that as the ice-sheets extended the sea-level fell, and as they retreated it rose again. At its lowest it may have fallen as much as

150 m below the present level. Such a drop combined with a colder climate, would have killed most of the coral reefs, and American geologists like Daly, studying the smaller reefs of the West Indies, were sure that these events had had an important effect throughout the world. The fall in sea-level would kill the corals and plane off the volcanic islands; as the ice melted the sea would rise again and many of the islands would be recolonised by reef corals, which would grow upwards keeping pace with the slow rise in sea-level.

It has recently become possible to test these theories, because a series of borings have been made into Pacific atolls. These have shown that many of the islands do have solid rock at the height expected by the glacial control theory of Daly, but some borings have penetrated thousands of feet of coral debris before reaching volcanic rock. It looks as if the continental margins have been more greatly affected by the glacial changes in sea-level than the islands in the middle

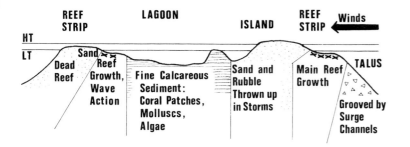

Figure 6.18 The coral reef environment. Notice how the zone of maximum growth is related to the area of maximum wave action. Draw a sketch-map of a circular coral atoll and plot the different types of sediment you would expect to find.

of the oceans. Both theories in fact can be true, and do not contradict one another as many have supposed. As more evidence comes to light we have to admit that both Darwin and Daly may have been right.

Below wave-base the finer muds settle out, or finer carbonates can accumulate. Even in this zone, however, there may be coarser debris, left by glaciers in time of lower sea-level, or due to the existence of local shell-banks. Thus grain-size is not conclusive evidence concerning the depth of water in which a sediment is formed.

2 Deep-water Deposits. Deep-water deposits (bathyal and abyssal environments) merge as the continental slope grades into the flat ocean floors. Much of the area is covered with the finest land-derived material, but the remains of planktonic life and volcanic dust may dominate areas sheltered from the entrance of mud and sand.

The turbidity currents, which are thought to be so important in the formation of submarine canyons, often move large quantities of debris rapidly, and some masses also slump down the continental slope when they become too thick and unstable on the edge of the continental shelf. Delta mouths are particularly susceptible to this process, and the ocean-floor south of the Ganges delta has a thick carpet of clastic, land-derived material which has reached it by this method. A flow of water, heavily charged with suspended sediment, moves downwards and spreads out over the abyssal plain, dropping the debris and helping to maintain the flatness.

Elsewhere the continuous 'rain' of skeletal remains from the surface plankton and swimming creatures is more rapid than the extremely slow local accumulation of detrital sediments: fine oozes build up. The character of the oozes depends on the depth of water and the surface climate of the area in which they form. **Calcareous oozes** are found on the ocean floors down to 3500 m, and are formed of the skeletons of such microscopic animals as the Foraminifera (mainly globigerina) and pteropods, mixed with clay. **Globigerina ooze**, for instance, covers over 200 million km^2 of the ocean floor. Oozes composed of silica dominate the areas between 3800 m and 5000 m deep, the silica being less easily dissolved and therefore able to descend farther, though accumulation is even slower. **Radiolarian oozes** cover over 20 million km^2 of the tropical Pacific Ocean, and **diatomaceous oozes** are common in Polar regions. **Red clay** occurs at depths beneath 5000 m, and is formed of iron and manganese from volcanic and wind-blown dust, and nickel probably from meteoric particles. Some radiolaria and shark's teeth are also included, and the deposit forms at a rate of approximately 1 mm every thousand years. The fact that the oozes accumulate so slowly is now helping us to work out some of the historical details of our planet, since so much time is recorded in such a small thickness of sediment. The type of mineral used in the building of the tiny skeletons is often diagnostic of the ocean surface temperatures when they were alive, and we can therefore find out what the climate was like at that time.

The fossils composing the oozes on the floor of the Atlantic Ocean are no older than the Cretaceous period; the oldest in the Pacific is of Jurassic age. This is important evidence which we shall mention again when we consider the origin of the ocean basins in Chapter 14.

Movements of Sea-level

We must realise by now that the level of the sea has varied considerably in recent geological time, and that it is a fact of great importance in understanding the formation of our present landscapes. The features of coral islands, of our coasts and of river valleys, cannot be explained unless we know something about these changes.

The coastal features produced by a changing sea-level are very distinctive. A rise in sea-level relative to the land may be caused by the sea itself rising, or the land sinking: the effect is the same. It leads to the submergence of those landforms nearest to the coast, and especially to the drowning of the lower parts of the valleys. In a hilly area like the South-west Peninsula (Devon and Cornwall) the drowned river mouths are known as **rias**. Regions which have been glaciated have deeper valleys, and these become drowned **fjords** like those in Norway, the sea lochs of western Scotland and the deep inlets of British Columbia, southern Chile and South Island, New Zealand. In low-lying areas the drowning causes the formation of shallow **estuaries**, like those of the Solent, Thames and Humber. **Submerged forests** and peat below high-water marks are common around British coasts, and provide further evidence to suggest that the sea has risen recently.

A fall in sea-level will leave an emerged coast, and the typical features of a normal coastline will be found many feet above the present high-tide mark as a **raised beach**. Raised beaches are particularly well-developed on the islands off the west coast of Scotland, where they locally provide flat ground for farming. Their essential features are illustrated in Figure 6.19. By examining the ground for these features, an old shoreline can be traced round the margins of the Thames basin, along the South Downs and into the Salisbury Plain. This ancient shoreline is the main piece of evidence for suggesting that the sea once rose to a height of 200 m above the present level (Figures 5.8 and 5.10), and it can be found again in the Welsh Peninsula.

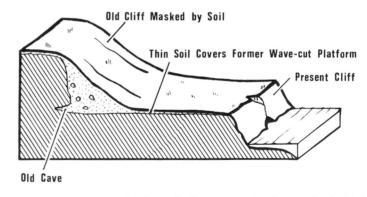

Figure 6.19 A raised beach. The main features are the flat area backed inland by a steeper rise or bluff.

A typical Norwegian fjord. What are the main features of the valley sides, the course of the fjord and the mountain-tops? (*Widerøe's Flyveselskap, Oslo*)

70

A Ria: Helford river, Cornwall. What effect has the rising sea-level had on the landforms? (*Aerofilms*)

Why has the sea-level changed so much over the last 2 million years? We know, and can understand, that the events of the Ice Age were certainly responsible for many of the changes. The sea-level fell as water was taken out of the oceans and frozen to build up the great ice masses which covered the northern continents, and it rose again as this water was returned on the melting of the ice. Another result of the masses of ice forming on certain areas of the Earth's crust was that their weight caused the crust to sag. When these areas were relieved of their ice burden it was some time before the land returned to its original position, and there was time for beaches to be carved by the sea. World-wide changes in sea-level, such as those caused by the melting of the ice sheets, are known as **eustatic**; those caused by local loading of the crust are known as **isostatic**. (Isostasy is discussed in Chapter 13.)

The Ice Age fluctuations of sea-level were very important, but in fact they were only an extra complication which affected a world-wide fall in sea-level. This began after the sea had reached the 200 m stage, and has continued to the present day. It is thought that it may be due to the formation of the great ocean trenches. As these features developed and deepened more and more of the ocean waters would flow into them, and the general level would fall. One geologist has calculated that enough water would be absorbed by this process to account for a lowering of 200 m all round the world.

Raised beaches on the Isle of Islay. What features shown here would lead you to believe that the sea-level had fallen twice? Can you see two sets of old cliffs? (*Aerofilms*)

7 Ice

One-tenth of the world's land surface is covered by ice today, and in the recent past great ice sheets extended over a large part of northern Europe and North America. Earlier periods of Earth history saw other ice invasions, though their record is less clear. We can therefore study both the areas where ice is active at the present time, and also regions which have just been uncovered after spending many thousands of years beneath ice.

How Is Glacier Ice Produced?

Snowfields become permanent above the **snow-line**. This is at a varying height depending on latitude and position with respect to the winds. At the Equator it is 6000 metres above sea-level, at 20° North and South it is 6500 m, and then declines to 600 m at 60° North and South: the highest points in Britain do not quite reach the snow-line, and so there is no permanent snow lying in these islands.

The extensive, deep covering of snow needed for the formation of glacier ice can only be supplied from very heavy precipitation in very cold areas. As the snow layer thickens it is changed to the more compact and granular **névé**, which increases in density as it is buried deeper and deeper in the snow mass. Eventually local melting, and the refreezing of percolating meltwater, will form a hard layer up to 30 m thick. **Ice** is formed as this hard-packed snow recrystallises, eliminating the air bubbles from the névé, and changing the colour from white to blue-grey.

How Does a Glacier Move?

A mass of ice and névé in a snowfield high up on a mountain slope is very unstable, and the weight of the increasing mound will eventually force the ice to move down into the valleys. The top ice acts as a brittle solid and fractures easily, but at depth it becomes viscous (ie flows stiffly, like pitch). Glaciers are characterised by deep crevasses in the brittle surface layer, some of them being 65 m deep. The weight of this thickness of rigid ice is too much for the underlying layer, which loses its solid properties and flows slowly downhill, carrying the surface layer along with it. Friction with the rocky sides causes the glacier margins to move more slowly than the centre. The amount of movement can be measured by inserting a line of stakes across the glacier between marked valley-side points, and returning some months later to ascertain the changes.

The speed of a glacier is governed by the slope of the valley it occupies, and the rate of snow supply. Thus the Alpine glaciers move 80 m per year, some in Greenland have been known to travel 30 m in a day and those on the continent of Antarctica, where very little new snow is provided, scarcely at all.

Glaciers end at a snout, where the melting processes become more rapid than the supply of ice. The position of this snout in the valley reflects the fluctuations of temperature or precipitation over a period of years: if it advances down the valley the air must be getting colder or the snowfall must be increasing; if it retreats our climate is becoming warmer or less snow is falling.

The Different Types of Ice Mass

Glaciation has two main manifestations. **Continental glaciation** is the complete covering of a landmass by an ice sheet several thousands of metres thick. The underlying rocks are only visible around the margins. Greenland and Antarctica are experiencing this today. **Mountain glaciation** occurs when ice is confined to the hollows and valleys in upland areas. The valley glaciers may spread out on surrounding plains if the climate is very cold. The Alps, the Himalayas and other high regions of the world, have short valley glaciers at the present day.

Corrie glaciers are developments from small snowbanks in sheltered

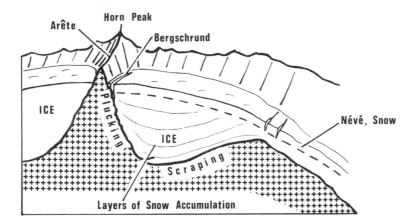

Figure 7.1 A corrie glacier. The weight of the ice causes the mass to move forward and open up a gap at the back of the amphitheatre. This gap is known as a bergschrund, and it becomes the drain for meltwater, which refreezes around the back wall rocks. When the ice moves forward again large fragments are plucked away with the ice, and these pieces of rock are used to scrape out the lower portions of the hollow.

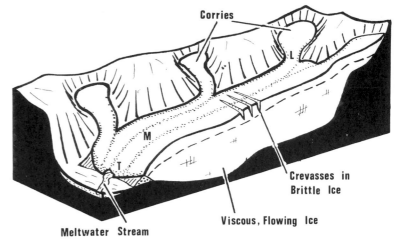

Figure 7.2 A valley glacier. The glaciated valley shows the contrast between the amount of the valley directly affected by ice and by a stream or running water (Figure 7.4). L, M and T are three types of moraine on the glacier surface—lateral, medial and terminal.

hollows which face away from the sun's rays and are therefore protected from too much summer melting. Daily thaw and overnight freezing break up the surrounding rocks, and the hollow becomes enlarged into an 'arm-chair' shape. The features of corrie glaciers are shown in Figure 7.1. They enlarge their hollow by wearing away the back wall and floor. In time two such corries may be separated from each other only by a steep, narrow, knife-edge ridge (an **arête**), whilst pyramid-shaped **horn peaks** (eg Snowdon and the Matterhorn) are formed at the junction of the arêtes.

Valley glaciers occupy former river valleys or newly formed glacial troughs, though they fill the whole valley rather than just a narrow bed in the bottom, as Figure 7.2 illustrates. Any unevenness in the valley floor will be reflected in the surface crevasses of the glacier. Glacial meltwaters contain a great deal of finely comminuted rock debris which gives the streams issuing from the snout of a glacier a milky appearance.

Piedmont glaciers are formed when glaciers coalesce as they leave mountain valleys and flow over a flat plain. The ice movement is slowed, and the ice itself becomes much thinner, often choked with rock debris. The Malaspina glacier in Alaska is the best example today, but in the past the Alpine glaciers flowed down their valleys and out on the surrounding plains of central Switzerland, Bavaria and northern Italy.

Ice caps are local masses of ice which cover a small upland area and discharge into valley glaciers. They are found in Iceland, Norway and Spitzbergen.

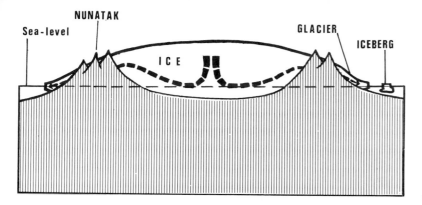

Figure 7.3 An ice sheet. This shows a generalised section across Greenland, where the weight of the 3000 m thickness of ice has caused the central area to be depressed below sea-level. As the ice moves towards the margins under pressure it is channelled between the coastal mountains, and the resultant valley glaciers move very rapidly.

Ice sheets are the largest form of ice mass, and develop as ice caps get bigger and bury a whole continent. The ice which covers Greenland (Figure 7.3) spreads out from the highest points of the great flat-topped dome it has built, and reaches the sea by fast-flowing valley glaciers. These have been pushed through the narrow mountain passes between the **nunatak** peaks protruding through the ice near the coast. Crevasses in the glacier ice cause it to 'calve' **icebergs** on reaching the sea. The Antarctic ice sheet is eight times larger, but the thickness of ice is more variable.

Most ice masses are shrinking at the present time. This indicates that our climate is becoming slightly warmer. Paintings of the Rhône glacier in Switzerland 100 years ago show it nearly 2 km farther down the valley than it is now. If all the ice left in the world melted now, the sea-level would rise 60 m.

Ice Transport

It has been suggested that the most important job a glacier does is to carry away the debris which falls on it from the frost-shattered peaks above, and which is worn off the rocks over which it passes. The great volume of ice in a glacier is able to hold a vast quantity of rock fragments of all sizes, and the thickness of glacial deposits in areas which have recently been glaciated testifies to this powerful function.

When boulders and finer material fall on to a glacier, the pile forms a **moraine** along the side (Figure 7.2): if the glacier moves slowly enough trees may even take root and grow on the deposit. At the point where two glaciers join two lateral moraines become a single medial moraine. Other debris works its way via crevasses into the glacier, and there is always a thick deposit of terminal moraine at the glacier snout, though the rushing streams of meltwater soon remove the finer particles.

Erosion by Ice Masses

A glacier can be likened to a great flexible file as it winds down a valley, smoothing, scratching and polishing the rocks. The larger ice sheets carry out most erosion in the central parts of the areas they cover, carrying away the loose soils, scraping the underlying rocks and leaving expanses of hummocky bare rock. Northern Canada, Sweden and Finland are like this, and some of the results can be seen in the far north-west of Scotland.

Angular rocks held at the base of the ice form deep scratches (**striations**) as they move across the bare rock, and when these are later exposed they help us to work out the former direction of ice flow. Projecting crags of harder rock are smoothed and plucked as the ice passes over them, and **roches moutonnées** result; soft rocks in the lee of hard **crag** may escape erosion and form a '**tail**', the best example of which is the Royal Mile slope leading down from Edinburgh Castle Rock.

The most dramatic effect of glacial erosion is seen where the youthful river valleys of mountainous areas have been occupied by ice. Uneven downcutting by the glacier leaves **rock steps** and deep hollows, which soon become filled by **ribbon lakes** as the ice retreats. Hollows of this type may be associated with a zone of softer rock or greater

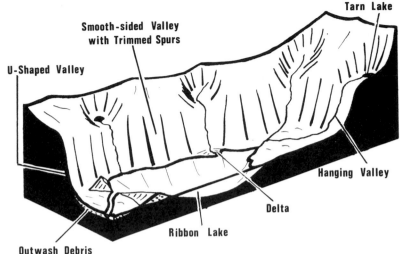

Figure 7.4 Glacial landforms. Relate these features to Figure 7.2.

weakness, or may occur where a tributary glacier has added a great quantity of ice to the main glacier and has thus increased its erosive power. The mass of ice and rock fragments moving along a valley trims off any spurs projecting across the valley, and carves out a wide, **smooth-sided trough** with a cross-section like an open U. The amount of erosive work done by a glacier depends on its speed of movement and thickness, ie upon the amount of new snow supplied each year. Tributary glaciers are smaller and not so thick, and do not erode their valleys so deeply. The over-deepened main trough is carved well below the old level of the river valley, which the tributaries joined without a

Edinburgh Castle. The craggy rock beneath the castle resisted the path of the ice across it and protected a 'tail' of softer rock behind—now the Royal Mile. (*Aerofilms*)

drop, but these same tributary valleys are now left hanging above the main valley. Waterfalls are common features where these **hanging side valleys** enter the main trough. Many of these landforms are illustrated in Figure 7.4.

During the last Ice Age some of the most active glaciers, in regions like Norway, western Scotland and South Island, New Zealand, carved very deep valleys as they flowed down steep slopes to the sea. The heavy precipitation on these mountainous coasts caused the glaciers to be very powerful and to carve deep vertical-sided gashes. The melting of the ice and subsequent rise in sea-level led to the drowning of these **fjord** troughs, which became long inlets of the sea.

Glacial Deposits

The debris carried by the glacier, or ice-sheet, is dropped as the glacier melts. Mountain valleys are so dominated by the majestic, steep erosional features, that the smaller-scale depositional features are often noticed less. In any case, such features as lateral moraines tend to merge with screes on the valley sides. Often only the **terminal moraine** ridges, crossing the valley, are important. Most of the lakes in northern Italy (eg Lakes Como, Garda and Maggiore) are dammed by terminal moraines at the mouths of the Alpine valleys, and the Great Lakes of North America were partially trapped by similar means.

Lowland areas once covered by ice-sheets now have a mantle of **boulder clay**, or till, of varying thickness and typically unsorted, containing a range of fragments from cobbles and boulders to the finest clay. The composition of this deposit also varies, depending on where the material originated, and it may be sandy, chalky or a stiff clay. Rivers, or the wind, working on these deposits since the ice departed may have sorted the different grades of rock fragment more thoroughly. Boulder clay plains are generally featureless and monotonous, but some of the small landforms often give rise to a little relief. **Drumlins** are elongated, oval mounds, with a steeper end facing into the direction of ice flow. They often occur in swarms in lowland areas, such as central Ireland, and the general aspect of such scenery has earned it the name 'basket of eggs' topography. The origin of these

drumlins is still a point for debate. They may be accumulations around a rock core, or may have been formed by stagnant pockets of ice near the margins of a retreating ice sheet, but are agreed to be sub-glacial. Some of the other distinctive landforms of glacial deposition are shown on Figure 6.2, and include the ridges of coarser material dropped by sub-glacial streams flowing into lakes of meltwater: these are **eskers**.

Ice may carry large pieces of rock weighing many tonnes from one area to another and then dump them as **erratic blocks**. If they are sufficiently distinctive to allow their source to be located they can help to determine the direction of ice movement. They could not have been moved by any other natural agency.

This layer of glacial drift tends to mask the underlying relief of solid rocks, and forms a completely new surface upon which rivers begin a drainage system when the ice melts.

The ice masses also have an important influence on the deposits in the area immediately in front of them. Lakes of meltwater are the site of **varved clay** deposition (Figure 6.2) whilst streams of meltwater drop debris to form terraces, and spreads of gravel and sand known as **kames**.

Tundra Areas

The very cold areas of the world, which are not actually covered by glaciers, are sometimes known as **periglacial** regions. They occur today in areas of tundra climate like Alaska, northern Canada and Siberia, where the temperatures are below freezing point for much of the year, and the bitter winds make it impossible for many plants to grow. When the surface soil thaws in summer the moisture it contains cannot sink downwards because the subsoil is permanently frozen. The water is thus trapped in the surface layers, which become very mobile, often flowing down a slope and carrying rock fragments. Since rivers seldom flow in these zones (being frozen for most of the year), the main processes at work are weathering by frost action, solifluction (flowing soil) and wind transport. These lead to rounded, hummocky landscapes.

The Quaternary Ice Age

Just over a million years ago ice advanced over much of North America and Europe, and glaciers formed in the mountain ranges and flowed out on the surrounding plains. Great ice sheets covered the northern parts of these continents, but they fluctuated greatly: there were long, warmer **interglacial periods** between the glacial advances, and from studies of tree pollens buried in peat bogs we can tell that the climate was warmer during some of these than it is now. The longest interglacial period was 300 000 years long and, as the last ice disappeared from Europe 10 000 years ago, we are probably living in an interglacial period now.

The Ice Age had a much wider effect. Whilst the northern landmasses were covered by ice all the climatic zones moved nearer to the Equator. Much of the Sahara desert became a rainy zone, and rivers scored the hillsides with valleys that can still be seen.

The Effects of the Ice Age on British Scenery

The British Isles were very much involved in this advance and retreat of ice, and many of our landforms have resulted from it. There seem to have been three major advances and retreats and minor fluctuations within them.

Most snow fell on the regions which at present have the highest rainfall. These include the mountains of north Wales, the Lake District and western Scotland. They must have had the thickest ice, and the most rapidly moving glaciers, and they certainly show the greatest effects of glacial erosion. All these regions have deep, U-shaped, trough-like valleys, such as the Nant Francon in north Wales and Glen Sannox on the Isle of Arran. The valley-heads are dominated by steep-walled corries, many of which are separated by arête ridges. The

Some glacial landforms. Make your own annotated sketches of these features. The top left-hand photograph is of Loch Einach; the others are in the Swiss Alps. (*Aerofilms and Swissair*)

Snowdon 'knife-edge', and Striding Edge east of Helvellyn in the Lake District, are two well-known examples of arêtes. The corries along the Nant Francon Valley are characteristically on the north-east facing sides of the ridges. Hanging valleys, striated rocks and valley-floor terminal moraines are other common features of these areas, but the lakes (lochs and llyns in Scotland and Wales) along the valley floors add most to the distinctive features associated with glacial erosion. Llyn Gwynant in Snowdonia, Thirlmere and Ullswater in the Lake District, and many Scottish lochs, testify to the uneven nature of the valley floors. Some of the deepest parts of these lakes are well below the present sea-level: Loch Morar in north-west Scotland has its surface 10 m above sea-level, but is over 300 m deep.

All of Britain north of the Bristol Channel-to-Thames estuary line was covered by ice at its farthest extent (Figure 16.10) though areas like the North Yorkshire Moors show typical periglacial features, and escaped burial by the later ice, which did not extend so far south. The South Downs on the other hand may have been capped by ice at one stage, although they were beyond the ice-sheet limit. The highlands of the eastern part of the country were affected to a much smaller extent than those in the west because of their smaller precipitation, and although great quantities of glacial debris were deposited in the lowland areas the ice produced few outstanding features. The deposits are over 30 m thick in places where they form the soft cliffs of the Yorkshire coast south of Flamborough Head, which has been pushed back 9 km since Roman times. Most of the soils in the lowland areas are formed on the glacial drift. Drumlins are especially common in the Vale of Eden in Cumberland, and in many of the Yorkshire Pennine dales. York stands on a notable terminal moraine left by one of the later advances of the ice.

As the ice retreated northwards the meltwaters were often trapped behind rock barriers, and when they overflowed they carved narrow, steep-sided valleys through ranges of hills. The River Severn, which once flowed east to join the Trent system, was diverted via the Ironbridge gorge into its present southerly course, and the Yorkshire Derwent, draining the eastern Moors, was turned inland to join the

Humber drainage because the sea ice blocked its former outlet at Scarborough. The North Yorkshire Moors are an excellent area for the study of these overflow channels: Newtondale is the most famous and striking since it is a deep channel used by a railway line, but not by rivers in its central course. Figure 7.5 summarises some of these features.

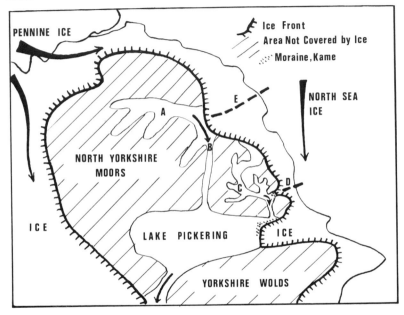

Figure 7.5 Ice and North Yorkshire. Notice the two directions of ice source. A, B and C are meltwater lakes and overflow channels; E is the course of the River Esk, which was resumed when the ice melted; D is the former course of the River Derwent, which was permanently diverted southwards by the ice.

Erratic blocks are common on the moorlands of the northern Pennines and in the boulder clays along the Yorkshire and East Anglian coasts, where ice from three directions met: pebbles of Shap granite from the Lake District, of rocks from the Cheviot area on the Border, and rhomb-porphyry from Norway can be found. Some erratics have even been discovered on local raised beaches in north Devon, an area which was beyond the ice sheet's furthest extent. It is possible that these erratic blocks, which include igneous fragments from Scotland, were contained in icebergs which drifted against the Devon coast and melted.

Although Britain is no longer covered by ice, the recent nature of that 'invasion' has left its distinctive marks freshly etched on our landscape. Glacial landforms are more noticeable in some areas than others, but every part of the country has been affected in some way. Even the southernmost parts show evidence of the tundra conditions and local ice-caps which prevailed only 10 000 years ago, and river terraces along river valleys like the Thames record the progress of sea-level fluctuations.

The Causes of Ice Ages

Why did the ice advance and cover much of North America and Europe? Why did it fluctuate with advances and retreats during that period? We are beginning to find some of the answers to these problems.

One of the best theories suggests that the changes began as great ranges of fold mountains were uplifted during the Tertiary era to isolate the whole Arctic area from winds bringing warmth from the Equatorial regions. Previously the world's climate had shown smaller variations in temperature, and Britain, for instance, had a sub-tropical climate at the beginning of the Tertiary: fossils of rhinoceros and crocodile testify to this.

The cooling of the Arctic area followed, and this affected the ocean waters of that zone. Masses of cold water began to move southwards and lowered the temperatures of the high latitude oceans. Since the ocean temperatures control the warmth of the air blowing on to the continents, temperatures began to fall over the land masses affected by these winds.

At the end of the Tertiary era one of the last major mountain uplifts closed the gap between North and South America. The warm surface waters of the Equatorial oceans were diverted northwards as the warm current we now know as the Gulf Stream (Figure 6.7). This led to greater evaporation from the oceans, and increased precipitation over the cooling lands. When this fell as snow rather than as rain the ice-sheets began to form. At first the Sun's rays were reflected by the ice and were not able to heat up the air above it: strong cold winds blew outwards from the ice masses and the ice advanced towards the seas. As the ice met the sea the contrast in temperatures increased the storminess and the ice-sheets increased in volume, whilst the sea-level was lowered considerably. This further intensified the bitter cold, but in so doing the processes of evaporation and precipitation were slowed down as the oceans began to freeze over. The ice-sheets, deprived of their snow supplies, began to flatten out, but still maintained the same areal coverage for some time. At last, however, the melting began to gain and vast quantities of icy water were poured into the oceans. The warming process took thousands of years, but eventually the oceans got warmer until the Gulf Stream began to take warmth northwards again at the end of the interglacial period.

This explanation accounts for many of the problems, but is not completely satisfactory. Further work will no doubt help us to understand the causes more fully.

THINGS TO DO AND DISCUSS, CHAPTERS 4–7

Investigation 1. Find out about your local water supply
Water boards are often happy to help with such inquiries. You could ask some of the following questions:

(*a*) How much of the water comes from underground sources?

(*b*) What is the geological situation (ie the aquifer and its characteristics; the nature of the structural trap for the water)?

(*c*) What is the local rainfall? How much of this: (i) runs off over the surface; (ii) percolates into the ground; (iii) is evaporated?

(*d*) If the water is stored in a reservoir, has a knowledge of the underlying geology determined its construction?

(*e*) Compare the processes necessary before (i) river water and (ii) underground water can be used by human beings.

(*f*) What is the difference between hard and soft water? Which types of area give rise to each of these characteristics?

Investigation 2. Measuring geological processes
A number of simple measurements are possible.

(*a*) *River flow*. You need to measure the cross-profile of the river and calculate its area (roughly depth × width). The velocity can be assessed by timing the passage of a floating object along 50 m of straight river. The amount of water passing in a certain time can be calculated using the following simple formula:

$$Q = V \times \text{stream profile area}$$
$$\text{(discharge)} \quad \text{(velocity)}$$

Carry out these measurements after heavy rain and after a dry period, and compare your results with the local (or school) rainfall records. If you can take a series of readings after a storm you will see how long the river takes to rise and fall. What causes the delay?

(*b*) *River load*. Remove a bottleful of water on each of the occasions when you measure the river flow, and allow the sediment to settle. The results can be even more worth while if you take bottlefuls from varying measured depths in the stream. The chemists may like to analyse the soluble fraction of the load; the suspended load which has settled out can be dried and weighed, and compared with the volume of water from which it was taken. Calculations can then be made concerning the amount of sediment transported in suspension and solution at the moment when the measurements were taken; an average for the year can be worked out allowing for the differences between periods of high and low water; and the yearly rate of erosion for the whole river basin can be assessed.

(*c*) Further information on the *rates of weathering and erosion* can be gained by examining fresh road cuttings and ploughed fields after heavy rain. The steepness of the road cuttings often leads to the formation of gullies, and the rate of enlargement can be measured. In many cases the road-constructors leave a bare cutting in soft rock to grade itself and allow the weathering processes to act on it. Once again the rate of change can be measured. You can devise your own means of doing this.

(d) If there is a lake near your home it is worth measuring the *rate of sedimentation* by keeping a close watch on the mouths of the streams emptying into the lake. Examine the type of material brought in and take samples from the bottom at various distances from the shore. Compare the amounts of sediment carried into the lake with that taken out of the lake by the outlet stream.

(e) If you live by the sea assess whether your *local coast* is one of erosion or deposition. Devise methods to measure the rate at which this is happening. What can man do to discourage erosion? What effects do his actions have on the sections of coast on either side of the area he has preserved? Historical information may be available in your local library.

(f) Has your local *river flooded* over its banks in recent years? Consult the local newspapers for that period to assess the extent of the damage caused. What was the cause of the flooding? What is the river authority doing to cut the risk of future flooding?

All these investigations will help you to increase your understanding of the processes at work and their rate of progress.

Investigation 3. Some laboratory experiments

Many natural processes are hidden from view beneath the ground, or take place on such a large scale that we cannot examine them and make simple measurements. Some can be simulated with simple laboratory equipment.

(a) *Porosity and permeability*. You will need a wide glass or plastic tube (diameter 3 cm), with one stoppered outlet; at least 100 cm³ of glass or plastic beads in each of three sizes (2 mm, 4 mm, 6 mm); a glass beaker; and a measuring cylinder. Place each set of glass beads in the tube in turn, and pour in sufficient water to cover the top level of the beads: measure and record the volume of water used, and this will give you a measurement of the porosity (ie the spaces between the beads, which represent rock particles). Then, also in each case, let out the water and record the time taken to empty the tube: this will give you the permeability (ie the rate at which a liquid will pass through

the sediment). Compare the data for the three sets of beads and draw up your conclusions (eg by a graph summary).

(b) A simulation of the behaviour of *density* (ie *turbidity*) *currents* can be carried out using the same wide tube. Fill this nearly to the top with water and tilt it at a low angle. Then pour into the top water containing a mixture of sand and mud. Record what happens to this mass and the way it is deposited at the bottom of the tube.

(c) *A delta model*. The equipment needed for this includes a large, round glass trough, a supported metal chute (eg a piece of aluminium held up by a retort stand); a rubber tube supplying water to the top of the chute; at least two varieties of sandy sediment, preferably in different colours and textures (stained if not); and a means of emptying the sediment into the chute. The water should flow continuously down the chute and into the glass trough, and some means of emptying (or controlling the overflow) should be provided. The different types of sediment are poured into the chute to build up alternating layers. Turn off the water supply when the sediment reaches the top of the trough; drain out the water and allow the sediments to dry. Cuts can then be made to demonstrate the internal structure of the delta.

Investigation 4

Make a chart to summarise all the new names of landforms which have been mentioned in Chapters 4–7, using the following headings:

Landforms found in regions of—

(a) Humid climate	(d) High relief/Tundra climate	(g) Shelf sea
(b) Limestone rocks	(e) Glacial conditions	(h) Deep ocean
(c) Desert	(f) Sea coast	

Practical book. Make a note of examples of these features you have seen. Give an accurate description of each landform and its location, aided by a diagram or sketch. Is it like the one described in this book? What are the differences?

Read 'Kraken Wakes' by John Wyndham. The later part of this science fiction story gives a good idea of what might happen if the world's remaining ice sheets melted. 'The Elements Rage' by F W Lane gives a lot of examples of the effects of natural forces on the land.

Some topics for discussion and essay-writing, based on O-Level questions.

(*a*) Define 'weathering of rocks'. Discuss the influence of climate upon weathering processes (L).

(*b*) Describe how a juvenile river differs from a mature river (O).

(*c*) Describe the effects of rejuvenation on a drainage system (O).

(*d*) Describe the influence of rock type and structure on the form of river valleys (AEB).

(*e*) Write an essay on the modes of formation of lakes (L).

(*f*) Describe the topographical features which result from deposition by rivers; mention briefly other features which might be present if the same lowland area had also been covered by ice (W).

(*g*) Describe the main types of sediments which accumulate under the sea. How may rocks formed in littoral, shallow water and deep water conditions be distinguished from one another (S)?

(*h*) Write brief notes on three of the following:

    (*i*) wave-cut platforms   (*iii*) spits and bars
    (*ii*) ria coastlines   (*iv*) submarine canyons (O)

(*i*) In studying a coastal region, what evidence would lead you to believe that there had been changes of sea-level in the past? (L).

(*j*) Describe the main physiographical features in Britain which have resulted from the Pleistocene Ice Age (O).

(*k*) Describe four features caused by glacial erosion in mountainous terrains and indicate how they were formed (AEB).

(*l*) What evidence in a formerly glaciated region would lead you to conclude that the erosive action of ice differed from that of water? (L).

(*m*) Describe, with diagrams, the following: escarpment and dip slope, alluvial flat, river terrace, raised beach, ria, and fjord. Explain briefly the origin of these features (C).

Part Three

Rocks and Fossils

8 The Sedimentary Rocks

The sedimentary rocks cover 75 per cent of the continental surfaces, and are thus the most accessible group of rocks for us to study. They are extremely important, because they are the only group of rocks to contain fossils, which are the main basis of the decisions we make as to whether a rock is older, younger or the same age as one in another region. We have already seen how many of the sediments are formed. Most of them are layers of debris deposited on the sea-floor, but some also accumulate on the land. We can often tell a lot about the rock and the conditions under which it was formed from its composition and the structures in it. Many different types of sediment are formed at any one period of time. On the land there are desert sands, glacial moraines, river alluvium, lacustrine and lagoonal salts and muds; along the coast are deltaic and estuarine muds, and beach gravels and sands; farther out to sea there are the shelf deposits in all their variety, and the finest oozes on the floors of the deep oceans. Each environment has a distinctive sediment associated with it, and different types of rock formed at the same time are often known as different **facies**, because the rock type has been related to the probable conditions of its formation—eg coral reef facies, deltaic facies.

From Soft Sediments to Hard Rocks

So far we have traced the geological story from the destruction of rocks by weathering and erosion, through the stages of transport and deposition, and have seen how sediments are forming today. Another important step, known as **lithification** or **diagenesis**, is necessary before a soft sediment becomes a hard, resistant rock. A series of

Fossils in a fine-grained mudstone. (*Crown Copyright*)

changes affect the sediment at conditions of low temperature and moderate pressure, and cause alterations over a long period of time.

Sediments often contain a high proportion of water, some muds holding as much as two-thirds of their volume, including a good percentage of dissolved matter. As other sediments are formed on top the underlying layers are compressed, and this water is squeezed out. The individual grains are packed firmly together until a very compact deposit results having a much lower porosity. The change from mud to mudstone may result in a decrease in thickness to one-tenth of the original, whereas sand may hardly change its thickness in becoming a sandstone (Figure 8.1).

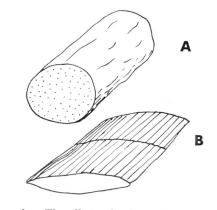

Figure 8.1 Compaction. The effects of sedimentary compaction on fossil tree stems preserved in sandstone A, and in shale B.

The material between the grains is known as the **matrix** of the deposit. The chemicals included in the water of the original sediment are often left behind, and may help to weld or cement the grains or shells together. In some cases the cement is introduced later, and is deposited in the pore-spaces. The chief types of cement in sedimentary rocks are calcite, silica and compounds of iron. Another process which takes place at this stage is replacement of the original minerals in the rock, especially if that rock was highly soluble. Seawater percolating through a porous limestone will dolomitise it by replacing the calcite with magnesium-rich dolomite.

The processes of de-watering, compaction, cementation and replacement gradually cause the rock to become hard. They may continue until the time when the rock in its turn is attacked by the atmosphere and broken up. We often find that the oldest rocks are extremely hard, whilst the younger ones are soft and poorly cemented.

Sedimentary Rocks: Evidence of the Past

The ways in which sediments are deposited and then altered to sedimentary rocks reflects the processes and the environment in which they were formed. An examination of sedimentary rocks will give us much information concerning these factors. There are certain features for which we should be looking.

(1) The most distinctive feature of sedimentary rocks is the fact that they are normally arranged in layers, known as beds or **strata**. Almost all of these layers were formed in the horizontal position, although sometimes scree may be formed at an angle of 30 degrees to the horizontal. Each bed of rock is bounded at the top and bottom by **bedding planes** (Figure 8.2), which mark breaks in the deposition and slight changes in the general conditions. The top surface of each bed is often hardened and may be marked by mudcracks, ripple-marks, fossil shells in their living positions and even raindrop pits. It may also be turned over by animals burrowing through the surface sediment in search of food particles. Very fine bedding divisions are known as laminations.

The layers of sedimentary rocks are not always regular in appearance. Some may be strikingly lens-shaped; this usually indicates a highly variable environment of deposition, such as that at the mouth of a river, or on a beach (Figure 6.15). **Graded bedding** is formed when a whole mass of debris is deposited rapidly, and the heavier fragments sink to the bottom whilst finer particles remain suspended in the water for some time before settling: it is often associated with turbidity current conditions. **Cross-bedding**, or false-bedding, is formed within a bed of rock and indicates the effect of wind or water currents on the movement and deposition of debris. A mass of grains is carried along as a surface carpet, tumbling over each other and rolling or avalanching down the front of an advancing mass of sediment. Greater thick-

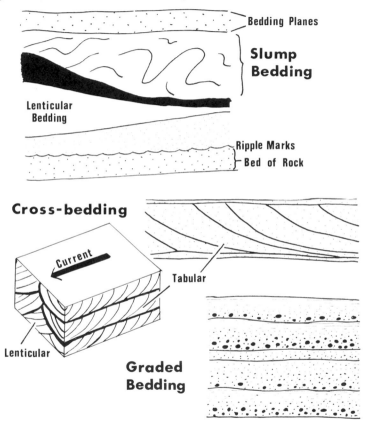

Figure 8.2 Bedding structures. Explain how these different types of bedding structure are formed.

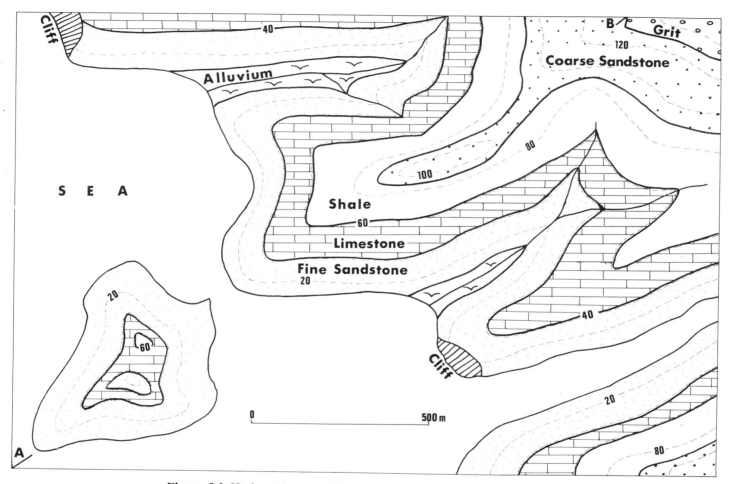

Figure 8.3 Horizontal rocks. All measurements in metres. Calculate the thicknesses of the limestone, shale and coarse sandstone. Why can you not work out the thicknesses of the fine sandstone and the grit?

nesses reflect the depth of water in which they were formed, but the thickest cross-bedded units are thought to have been formed by the wind. This internal structure is called false-bedding because it sometimes occurs on such a large scale that the dips of the internal laminae may obscure the true dip of the bed as a whole. **Slump bedding** occurs when unconsolidated sea-floor deposits are disturbed and crumpled, often due to collapse down a steep slope.

(2) Within each bed of rock the **sizes, shapes and arrangements of the individual grains** also tell us important facts. The grain size is one factor often used to divide different groups of rocks, although some sediments contain grains of widely varying sizes (eg boulder clay). We say that such sediments are poorly sorted: they were formed too rapidly to allow sorting to take place. In a well-sorted deposit (cf Figure 8.4) all the larger fragments would have been dropped or broken into smaller fragments, and the finer particles winnowed away, leaving all the debris of a similar size.

Angular grains indicate that they have not been transported very far, and rounded fragments are evidence of considerable travel (though roundness also depends on the size and composition of the fragments involved); wind deposits are characterised by rounded, polished sand grains and faceted pebbles. On the other hand oolitic limestones with their characteristic egg-like structure are commonly found in warm, shallow seas. Thus many tiny details have to be investigated to enable us to make the most of our rock studies.

(3) The **composition** of the sediment is also related closely to the conditions in the source area from which it is derived (ie parent rock, weathering processes, speed of erosion), and the conditions at the site of sedimentation. A river may be eroding an ancient desert sandstone, and the sediment it deposits in the sea will have a strong resemblance to one formed by the wind because it is still formed of the same wind-polished, 'millet-seed' grains which have been altered little by river transport. Minerals in a rock can often be traced back to a distinctive source region: thus the Millstone Grit rocks of the southern Pennines contain unusual minerals which suggest that they have been derived from rocks similar to those at the surface today in northern Scotland and Norway.

Rapid erosion and deposition will produce a sediment with a wide variety of rock and mineral fragments in terms of composition and size, whilst slower processes will sort out the detritus, leaving behind the larger and heavier materials. Strong currents in the area of deposition will keep all but the coarsest material moving. It is helpful to think of high energy environments with strong currents, in which only material of sand grade and above is deposited, and of low energy environments where the stillness enables mud and dust to fall out of suspension (Figure 6.14).

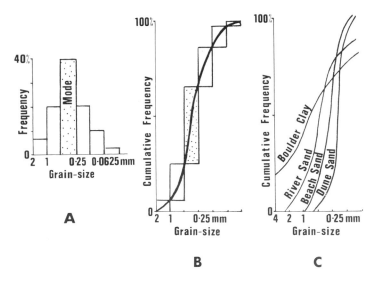

Figure 8.4 Sorting of sediments by grain-size. A shows the result of plotting an analysis of a sediment after sieving it to separate the grains of different sizes: this is a simple histogram. In B this is transferred to a Cumulative Frequency graph, and a curve drawn. C shows the characteristic curves for sediments formed under particular conditions. NB The horizontal scale is the same in each case.

Classifying and Identifying Sedimentary Rocks

Classifications are designed to mean something to those who use them. Several methods of classifying sedimentary rocks are available, but the best one for our purposes is that which can be easily related to our field- and hand-specimen observations.

Figure 8.5 outlines the classification we shall use here, and yet another approach is suggested at the end of this chapter. As with all rock classifications we are picking out types, between which there are many intermediate varieties: there are very few 'pure' sediments like chalk (99 per cent calcium carbonate). The major component determines the name (eg calcareous sandstone, muddy limestone).

We make a basic division between the sediments which are largely composed of material which has been transported before deposition—the fragmental sediments—and those which are precipitated directly on the sea-floor. The fragmental components come from erosion, weathering and transport on the land (clastic), or from the accumulation of largely carbonate material from shell-bearing and skeleton-forming creatures (bioclastic). If erosion and transport are rapid on the land the clastic material poured into the sea dominates all the other sediments; if the land is low-lying and little clastic material reaches the sea, the slower bioclastic accumulation results in lime-rich rocks.

ORIGIN OF MOST IMPORTANT COMPONENTS		ROCK-TYPES	Common names
TRANSPORTED MATERIALS	CLASTIC	RUDITES (PSEPHITES)	Conglomerates, breccias
		ARENITES (PSAMMITES)	Sandstones, greywackes, arkoses
		LUTITES (PELITES)	Mudstones, shales
	BIOCLASTIC (mainly carbonates)	CALCIRUDITES CALCARENITES (+Oolites) CALCILUTITES	Bioclastic limestones
		Some COALS	Cannel coal
UNTRANSPORTED MATERIALS	BIOGENIC	REEF LIMESTONES FORAMINIFERAL, COCCOLITH LIMESTONES OOZES—Calcareous —Siliceous	Biogenic limestones
		Most COALS PHOSPHATES	
	CHEMICAL PRECIPITATION	LITHOGRAPHIC LIMESTONES DOLOMITE, MAGNESIAN LIMESTONE IRONSTONES	Chemical limestones
		EVAPORITE SALTS	Rock salt, potash
		ALUMINOUS DEPOSITS	Bauxite
		Some SILICEOUS DEPOSITS	Chert, flint

Figure 8.5 A classification of sedimentary rocks.

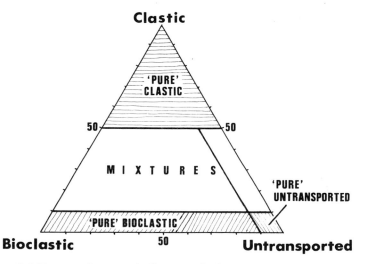

Figure 8.6 Ternary diagram. A diagram which can be used to classify sedimentary rocks, which are all essentially mixtures.

The sediments formed without horizontal transport are largely of biogenic or chemical origin. Organic action produces a variety of reef limestones, deposits of algal and foraminiferal oozes, most coals and phosphate rocks. Chemical rocks are formed by direct precipitation of iron or salts, or by later (secondary) chemical reactions in which the original materials are replaced. Thus dolomite replaces calcite in limestones and siliceous rocks like flint and chert form as nodular layers after water percolation, whilst aluminous rocks result from chemical weathering.

The following treatment of the main rock types recognises the overlapping nature of such a classification. After a consideration of the main clastic groups, the limestones are taken together, rather than treating each type separately as it occurs in the classification. The bioclastic and non-transported sediments are therefore treated according to their major chemical constituent. Three rock groups—shales, sandstones and limestones—make up 98–99 per cent of the crustal sedimentary rocks by volume.

Rudites: Conglomerates and Breccias

The fragmental group of rocks is divided according to the size of the grains they contain. When most of the grains are over 2 mm across the rocks are known as conglomerates if the fragments are rounded, and breccias if they are angular. American geologists use the descriptive term, sharpstones, when they refer to breccias. Within the whole group are rocks formed of large boulders, pebbles and small granules bound together by a cement which may be either a mineral like silica, or a matrix of fine mud.

Conglomerates are produced when the fragments have been rounded after transport, or after some time on a pebbly beach. We have all seen the results of coastal wave-action, which throws pebbles against each other and soon rounds off the corners: the same thing goes on in rivers though it is not as a rule so obvious. Some conglomerates are formed of one type of resistant pebbles like those of flint or quartzite, which have been well-rounded. They form thin layers in places, and record the slow encroachment of the sea over the land.

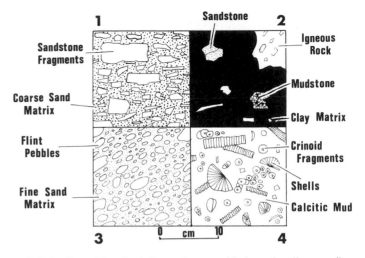

Figure 8.7 Rudites. Use the information provided on the diagram (ie scale, composition, roundness, etc) to suggest names for the rock-types depicted. How were they formed?

An example is the Hertfordshire 'Puddingstone' of Tertiary age, which contains small, round flint pebbles cemented by silica. Another type of conglomerate may be formed from a mixture of different rock fragments of all sorts of shapes and sizes, including pebbles of limestone and other rocks which are usually soon lost in transport. These conglomerates form thicker layers, and were the product of rapid mountain erosion, such as occurred in Britain during the early Old Red Sandstone period.

Boulder clay has been mentioned as a glacial deposit. It is usually a mixture of pebbles in a fine-grained matrix, though the actual composition depends on the source of the rock fragments contained: the chalky boulder clays of East Anglia are very different from the deposits formed from clay and sandstone rocks in the Midlands; the boulder clay of the Yorkshire coast has a few large boulders set in a chocolate-coloured matrix of clay, whilst the Old Red Sandstone rocks of Pembrokeshire gave rise to a reddish deposit with many more rock fragments. When a boulder clay is compacted it will harden to a **tillite** and the finer matrix may recrystallise as a cement.

Breccias are formed of angular fragments, and are usually found close to the scene of the weathering which led to their formation. Most breccias are composed of scree accumulating at the foot of a mountain slope, like the Brockrams of New Red Sandstone age in the Eden Valley of Cumberland. This deposit contains a proportion of rounded fragments, and is sometimes called a breccio-conglomerate. When a fault moves, the rocks on either side are often crushed or broken, but the fragments may be recemented without travelling at all.

Arenites: Sandstones, Arkoses and Greywackes

The members of this group, in which the main grain sizes are between 2 mm and $\frac{1}{16}$ mm, are often known as the **arenaceous** rocks. Most of them contain minerals and tiny rock fragments which can just be seen with the naked eye, and another characteristic feature of these rocks is that they have a rough feel when the surface is rubbed. The sandstones, and rocks of related grain size, are the largest group of sediments, but as in the case of the conglomerates the division of the fragmental rocks according to grain size masks several vital distinctions. Rocks within the group may be formed under very different conditions, though they all require active currents to prevent finer materials from settling: they are common in high energy environments. We recognise four main varieties:

(1) **Orthoquartzite** is a sandstone composed almost entirely of quartz. Rounded quartz grains are cemented together by silica, which often forms crystal-shaped outgrowths around the transported grains. These grains are usually very well-sorted, and, although cross-bedding and ripple-marks are common, fossils are rarely found in these rocks. They are called orthoquartzites to distinguish them from similar quartz-dominated rocks which have been metamorphosed (metaquartzites—

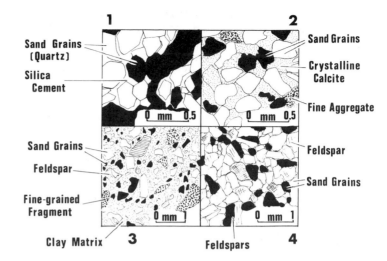

Figure 8.8 Arenites. Use the information provided on the diagram to suggest the way in which each rock was formed, and to identify their names. The drawings are taken from magnified thin sections of the rocks.

cf Chapter 11). Sedimentary quartzites occur as widespread deposits up to a hundred metres thick, and are often associated with layers of limestone and thin underlying conglomerates. The conditions in which they were formed are illustrated in Figure 8.9(A). In such conditions only the most resistant grains (ie quartz) survive, and the resultant rock is often extremely durable. One British example is the Stiperstones quartzite of Ordovician age, which forms a distinctive ridge in Shropshire.

Many **other sandstones** are also formed largely of quartz grains, but may have a different cementing material such as calcite, or an iron compound. Desert sandstones, for instance, have rounded millet-seed quartz grains coated with red or mauve iron hydrate which helps to cement them together, many of the rocks in the New Red Sandstone

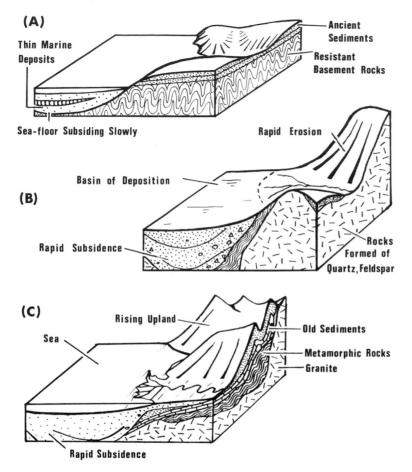

(A)

Thin Marine Deposits

Ancient Sediments

Resistant Basement Rocks

Sea-floor Subsiding Slowly

Rapid Erosion

Basin of Deposition

(B)

Rapid Subsidence

Rocks Formed of Quartz, Feldspar

(C)

Rising Upland

Sea

Old Sediments

Metamorphic Rocks

Granite

Rapid Subsidence

Figure 8.9 How are sandstones formed? The conditions in which orthoquartzites (A), arkoses (B) and greywackes (C) are formed. What are the differences in the types of source rocks, rates of erosion and conditions in the areas of sedimentation?

group being of this type. The true greensands of southern England are formed of quartz coated by one of the iron silicates, glauconite, and when this is removed by weathering the soils produced are very poor because they consist almost entirely of ash-like quartz sand. The mineral glauconite is interesting because it is formed only in marine conditions, and combined with the frequent occurrence of current-bedding in the greensands tells us that these rocks were formed in shallow seas, very much like those in which the orthoquartzites were formed, but with a greater proportion of iron salts entering the solution. Most sandstones contain a small proportion of distinctive heavy minerals, which help us to fix the origin of the sediment debris: deposits in Devon, for instance, containing tiny fragments of tourmaline, would have come from the erosion of the Dartmoor granite. When such heavy minerals increase in the proportion of the rock they occupy the whole deposit is known as a **placer**. Examples include the tin-bearing gravels of Malaya and the diamond-bearing beach deposits of South-west Africa: they are both recent deposits washed out of the original veins. The gold-bearing gravels of the Witwatersrand region in South Africa are placers in Precambrian clastic stream channels.

Another group of quartz-rich arenaceous rocks is found immediately beneath many coal seams. These **ganisters** are almost pure quartz, but contain a few black rootlets standing out from the white rock. The great coal forests, which used these deposits as a soil, removed all the mineral plant-foods and left only the quartz behind.

(2) **Arkoses** are sandstones containing a high proportion of feldspar minerals, and very little matrix. With a smaller proportion of feldspar rocks may be known as feldspathic sandstones, or **grits**, and many of these occur in the Millstone Grit of the Yorkshire Pennines, and in the Middle Jurassic rocks of the North Yorkshire Moors: in each of these cases the rocks were formed in deltaic conditions. True arkoses are nearly always associated with products of the erosion of granite and metamorphic rocks of similar composition forming high mountains (Figure 8.9(B)). Normally feldspar is soon decomposed by weathering in humid regions, and sediments containing a high proportion of this

mineral (over 25 per cent) must have been formed very rapidly, or in deserts. Arkoses usually contain a mixture of angular and moderately rounded, coarse and fine grains, and occur in thick wedge-shaped formations or in thin layers, depending on the rate of subsidence of the area in which they were laid down. They tend to be pinkish in colour because of the orthoclase feldspar, which is the variety most resistant to decomposition, and is therefore usually the dominant feldspar. The erosion of ancient mountains in north-west Britain led to the formation of Precambrian arkoses, which now form majestic mountains themselves rising above the levelled basement rocks along the west coast of northern Scotland. Thick arkosic deposits, known as **molasse**, were formed to the north and south of the Alps towards the end of the earth movements which resulted in their elevation.

(3) **The greywackes** form another contrast, since they have a high content of fine matrix material made up of clay minerals and tiny flakes of mica or chlorite. This engulfs the sand-size quartzes, rock fragments and feldspars which are characterised by poor sorting and angular shapes. Greywackes are formed in areas of instability where earth movements are common. Rapid erosion and transport are accompanied by rapid subsidence of the basin of deposition (Figure 8.9(C)), often known as a **geosyncline** (see Chapter 12 for a fuller description and discussion of such features), and the sediments accumulate very rapidly. Greywackes are common in central and northern Wales, the Southern Uplands of Scotland and the central part of Devonshire. They are greyish mottled rocks, weathering to rusty brown, and are associated with deep-water black shales, volcanic lavas and ashes, and radiolarian cherts. Graded bedding is often found in greywackes, and although the individual layers may be quite thin, the total thickness of greywacke in an area may be up to 10 000 metres.

(4) **Sub-greywackes** are perhaps the most common of all the arenaceous group, but are not so distinctive as the other types, since they are often a general mixture of minerals. They are mainly composed of quartz grains, plus a proportion of rock fragments and feldspars, but have a smaller amount of clay matrix and more mineral cement than true greywackes. Better sorting, and a greater degree of rounding, characterise the grains, but the colour is similar to that of a greywacke. They are common in this country in the Coal Measures, and many of the sandstones formed between the coal seams are of this type. Current-bedding and ripple-marks are common on the upper surfaces of the rock layers, and where there is a higher quantity of mica present the rocks split easily and are known as **flagstones**.

Lutites: Silts, Shales and Mudstones

The finest detritus and colloidal material includes the silts (between $\frac{1}{16}$ and $\frac{1}{256}$ mm) and muds (less than $\frac{1}{256}$ mm), and once again the constituents have varied origins. Fine 'rock-flour' comes from glacial scraping, new clay minerals have been produced by the chemical weathering of feldspars and ferromagnesian minerals, and other salts are precipitated where fresh water enters the sea. The silts have a higher proportion of quartz grains than the muds. The minerals in muds are so tiny that they can be distinguished only under very high-powered microscopes. Such fine-grained sediments can settle out only in low energy environments, which exist in currentless lagoons, lakes, or in the sea beneath wave-base. The individual grains cannot be seen when we examine these rocks in hand-specimen, and they have a very smooth surface. The finest-grained rocks are often known as the **argillaceous** group.

As the original mud deposit has the water squeezed out, it changes to a sticky clay, and, as the process continues, to a hard mudstone or laminated shale. Some special types will illustrate the variety of argillaceous rocks.

(1) **Black shales** are very common in the lower parts of the geosyncline basins that were later filled in by greywackes, and they are plentiful in Wales and the Southern Uplands. They have a high proportion of organic debris, and often contain nodules of pyrite (FeS_2) formed as the sulphur from dead organisms combined with iron salts in the deep, oxygen-free waters. Conditions for the formation of these

rocks must have been similar to those prevailing in the Black Sea at the present day (cf Figure 6.6).

(2) **Fireclays** are the fine-grained equivalent of ganisters. Both of these seat-earths occur beneath coal seams, and are very rich in alumina (38–44 per cent), which makes them of use in the building of steel furnaces. Fireclay is usually light grey in colour without any bedding structures, and is often crossed by carbonised rootlets.

(3) **Oil shales** are found in central Scotland. They are black or brown in colour, and give a curved fracture when broken. The hydrocarbons they contain were used for some years as a source of oil.

(4) **China clay** is composed almost entirely of the clay mineral kaolinite, which is formed in the late stages of the cooling of a granite mass, when percolating solutions and gases react with the feldspar minerals. Most of the British supplies come from the granite intrusion just north of St Austell in Cornwall, and they are shipped to the centre of the country's pottery and ceramic industry at Stoke-on-Trent.

(5) Other types of argillaceous deposit include **marl**, which has a high proportion of calcium carbonate; **Fuller's Earth**, which is very absorbent and is used in the refining of oils and fats; and the **Alum shales** found in the Jurassic rocks of the Yorkshire coast.

Bricks and **tiles** are produced from clays, mixed with a small proportion of sand to reduce shrinking in the firing process. The best brick-making clays are those containing natural fluxing materials like soda and lime, iron to give the red colour, and organic matter to help the firing. The Oxford Clay (Jurassic age) is one of the major sources in Britain, and is quarried extensively near Peterborough and Bedford. Most clay horizons have been used for brick-making at some stage, but the many abandoned brickpits testify to the centralisation of the modern industry. There are still small brickworks, such as the one at Hambledon in Surrey, producing bricks directly from the Weald Clay (Cretaceous) for special uses (eg to bear heavy loads, or for special facing effects). Tiles are made in a similar way, but increasing numbers are being made of concrete in moulds and being given a surface dressing for colour and appearance.

The Carbonate Rocks: Limestones

In most sediments there is a certain proportion of debris which is organic or chemical in origin, but it is only when these elements become dominant that they give rise to a distinctive rock. Such conditions are only found where the supply of clastic debris is reduced to a minimum, and non-clastic rocks are most likely to be produced in times of shallow, widespread seas and slow erosion. Whilst the clastic rocks may be of the greatest importance where quantity is concerned, it is the sediments composed of organic and chemical materials that are the most important economically.

The limestones embrace a large group of rocks of chemical and organic origin, but most of them have a percentage of clay or sand.

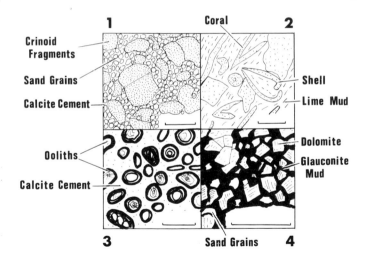

Figure 8.10 Carbonates. Use the information provided on the diagram to suggest the way in which each rock depicted was formed, and to put forward a name for them. The scale line is 1 mm long in each case. The drawings are taken from magnified thin sections of the rocks.

Most limestones are **bioclastic**, containing transported, organic debris rich in calcium carbonate, and grading into clastic rocks. Some are **biogenic**, representing an ancient reef structure, shell bank or ooze. A third group has a **chemical** origin and is often associated with the formation of evaporite salts. Mixtures are common.

In Britain we have a wide range of limestone types. Four of the most common are described here.

(1) **Chalk** is the purest type of limestone, containing a tiny proportion of sand or mud. It is white and brittle, and is mostly formed of the remains of algae and broken shells. One-eighth of the rock is usually made up of other tiny fossils, such as Globigerina, and there are also larger shells of sea urchins (echinoids), belemnites and lamellibranchs. (Details of all the different fossil groups will be given in Chapter 9.) The chalk of Britain forms escarpments in the south and east of England, and the nature of the rock varies so little throughout the 400 km of its outcrop that the conditions of formation must have been uniform over widespread areas. It is thought that the chalk was formed in relatively shallow waters surrounded by deserts, which would supply very little clastic debris to the sea.

(2) **The harder limestones of Carboniferous age** in Britain include a variety of calcareous (ie limy) deposits. There are **shelly limestones** where up to three-quarters of the rock is composed of the remains of sea lilies (crinoids), or corals and brachiopods. The rest of the rock is formed of calcitic mud and ordinary detrital mud. Most of these rocks are grey in colour and weathered surfaces often show the fossils to advantage, since the matrix around them has been destroyed first. Such limestones form thick, massive beds which are well-jointed. They are non-porous, but the joints provide easy routes through the rocks for water, which attacks and dissolves the calcium carbonate. This combination of strength and solubility leads to the formation of caverns and pot-holes. Chalk is too weak to support similar features.

Reef limestones are another type commonly found in the Lower Carboniferous rocks. They are formed of the hard skeletons of colo-

nial corals, of algae and of many other fossils which once existed near the living reefs. Such reefs form compact, rounded masses of hard, nodular limestone in the midst of well-bedded layers, as shown in Figure 8.11, and when erosion takes place they stand out as low hillocks or knolls, which are a feature of parts of the Pennines. The

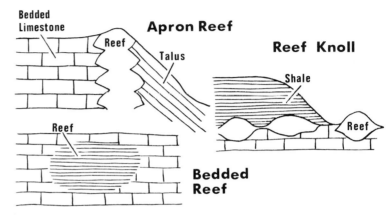

Figure 8.11 Reef limestones.

porosity of these reef limestones has made them important reservoir rocks for oil in many parts of the world.

(3) **The Cotswold Hills of Gloucestershire** are formed of thinner layers of oolitic limestone. These rocks are largely of chemical, shallow-water origin, in the manner already discussed in Chapter 6: wave action in highly calcareous seas gives rise to the fish-roe-like ooliths. Larger fossils may be included in the rocks.

(4) **Magnesian limestone** is an important component of the Permian rocks in north-eastern England. Much of the original calcium carbonate has been replaced by magnesium carbonate, and the rock is a buff-coloured, massive sediment which forms the North Sea cliffs of County Durham, and an escarpment in Yorkshire. When the processes

of replacement affect the rock completely all the fossils, reef structures or ooliths disappear, leaving rhomb-shaped crystals of dolomite ($CaCO_3.MgCO_3$).

We thus have many varieties of limestone in Britain. In other parts of the world deposits of **tufa** are found around the mouths of caves from which lime-rich streams emerge. The famous **lithographic stone** of central Germany was formed in the Jurassic age by the hardening of calcitic mud. The deposit is so fine that the tiniest details of fossils entombed in it are preserved, including the imprint of the feathers of the first bird known to man.

Limestones are the most-quarried rocks in Britain today. Their most important use is in the production of **cement**: two-thirds limestone and one-third clay are mixed in the manufacture. Some rocks, like the Lower Lias at Rugby, are natural cementstones because the lime and clay are already mixed before extraction. Chalk is the basis of 75 per cent of British cement, and the largest works in the country line the banks of the lower Thames, located where the Chalk outcrop comes close to the river muds (with which the chalk is mixed) and the transport artery. Other large works, such as that near Castleton in northern Derbyshire, also thrive on a site near the junction of limestone and shale.

The older and more resistant Devonian and Carboniferous limestones are also becoming increasingly used as a source of **roadstone**, and large groups of quarries near Plymouth and in Derbyshire quarry the rock for this purpose. Limestone bonds well with asphalt, and the motorway construction programme has stimulated this development. The steel industry also uses local limestones as a **flux** in its blast furnaces, and magnesian limestones of northern England are used in the making of **refractory bricks** for lining steel furnaces. The oolitic limestone is a handsome **building stone**, as towns like Bath, and many Cotswold villages, testify. Portland Stone was used for many of London's most impressive buildings (eg St Paul's Cathedral).

Many limestones, especially those of reef origin, are extremely porous and make important oil **reservoirs**. The process of dolomitisa-

tion reduces the rock volume within a certain space still further (by 12 per cent) and thus assists in increasing the porosity. Chalk is the most important aquifer in Britain (cf Chapter 4).

Ferruginous Rocks: Ironstones

Most of the world's supply of iron ore comes from sedimentary beds of haematite (iron oxide), limonite (hydroxide), siderite (carbonate) and chamosite (hydrated silicate), or from metamorphosed sediments. The British supplies of ironstone are now mostly imported, but there is an important band of low quality ore of Jurassic age running northwards through Northamptonshire to Scunthorpe near the Humber estuary and thence under the North Yorkshire Moors. It is a mixture of chamosite and siderite, with a little limonite and haematite, and often has an oolitic texture, suggesting that the iron compounds have

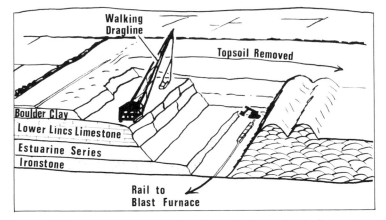

Figure 8.12 Opencast ironstone mine. Describe the method of operation depicted here. Compare the huge Walking Dragline with its boom up to 100 m long with the mechanical shovel tipping the ore into the railway trucks. Farming is interrupted for only a few years, and can be restarted as soon as the topsoil is placed on the levelled overburden which has been dumped behind the workings by the Dragline.

replaced an original limestone. This rock contains between 24 and 38 per cent of iron, and is mined by opencast methods and smelted close to the quarries (Figure 8.12). Siderite occurs associated with coal in the Carboniferous rocks as blackband ironstone, which in the nineteenth-century was the main source of iron ore in Britain.

The famous Lake Superior iron deposits of the United States of America were all formed from a sedimentary mixture of iron compounds known as taconite. This originally contained only 20 per cent iron, but leaching removed other soluble constituents and left an enriched haematite ore with 50 per cent iron. The recently discovered Labrador deposits have been similarly enriched.

The Non-fragmental Rocks: Salt Deposits

Extensive accumulations of salt are produced by the evaporation of

THE STASSFURT SALT DEPOSITS (PERMIAN AGE)

ROCK-TYPE	THICKNESS	CONDITIONS OF FORMATION
Shales, Sandstones, Clays		Wet conditions
Rock Salt (NaCl) Anhydrite ($CaSO_4$)	Variable 30–80 m	Dry conditions, intense evaporation
Salty Clay	5–10 m	Wet conditions
Carnallite ($KCl.MgCl.6H_2O$) + Kainite ($KCl.MgSO_4.3H_2O$) + Sylvinite (KCl) Kieserite ($MgSO_4.H_2O$) Polyhalite ($K_2SO_4.MgSO_4.2CaSO_4.2H_2O$)	16–40 m	The driest conditions: the most soluble salts were precipitated
Layered Rock Salt and Anhydrite	Hundreds of metres	Seasonal alternations in arid climate
Anhydrite, Gypsum		Dry conditions with evaporation

Figure 8.13

lake and sea waters in very dry climates. Some of the most famous deposits occur at Stassfurt in East Germany (Figure 8.13), but there are also important deposits in Britain near the Tees Valley (anhydrite and rock salt), and the Weaver Valley of Cheshire (rock salt). Potash deposits have been discovered beneath the North Yorkshire Moors, but are too deep to be mined economically at present (Figure 17.8).

(1) **Rock salt** (NaCl) forms massive layers coloured yellow or reddish tints by impurities. It is soft and light, and is usually mined by pumping water into the layer and extracting the brine solution. This often leads to subsidence of the landscape.

(2) **Anhydrite** ($CaSO_4$) often occurs as fibrous, granular or more compact masses with a grey, blue or reddish tint. It is harder and less soluble than rock salt, and is usually mined. The ICI mine at Billingham (Co. Durham) cuts into a layer of anhydrite with an average thickness of 6 m. Passages 16 m wide are cut out in a grid pattern, leaving 16 m square pillars in between: no other support is necessary.

(3) **Gypsum** ($CaSO_4.2H_2O$) is rather similar to anhydrite, but is much softer and is used in the making of plaster of Paris.

These are the three main salt deposits, and all are used to a great extent in the chemical industry, where they are made into a variety of products from fertilizers to plastics. Smaller quantities of other salts, including those mentioned in Figure 8.13, are also utilised.

The weakness of many of the salt deposits compared with the other sedimentary rocks causes them to flow under continuous pressures, even if these are relatively slight, instead of folding or fracturing. Salt is thus often concentrated into large, dome-shaped masses, which are common round the Gulf of Mexico and beneath the North Sea where they have become oil traps. Figure 8.14 shows how such salt domes form; look also at Figure 8.20.

There is one great unanswered question concerning the salt deposits: how did they accumulate to such great thicknesses in the past? If you evaporated seawater a thousand metres deep you would only leave a few metres of salt behind. Yet if you look at the diagram

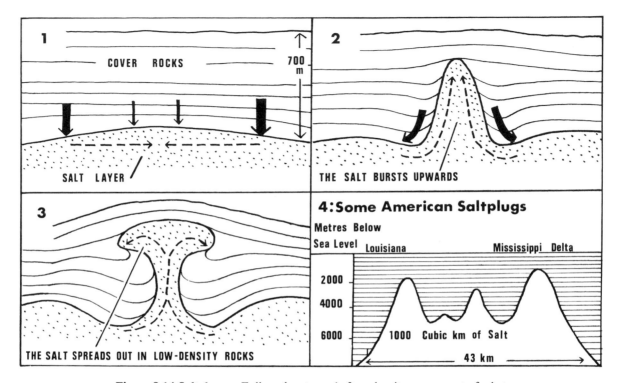

Figure 8.14 Salt domes. Follow the stages 1–3, and write an account of what happens in the formation of a salt dome, remembering that salt has a low density. The arrows in stage 1 are proportional to the pressure exerted on the salt layer at different points. Note the vast quantity of salt involved in the American domes.

of the Stassfurt deposits (Figure 8.13) you will see that they are too thick to have been formed in a simple phase of evaporation. The Wieliczlic mines in south-western Poland have become a tourist attraction because a whole industrial town has been carved underground in the salt layer which is 400 m thick. The best answer so far is that the area of salt deposition was subsiding causing a continual renewal of inflowing seawater, and that flowage of the mineral led to further concentration of the deposit. It is known that the Gulf Coast of the United States of America has subsided by many thousands of metres, and that some of the great salt domes in this area have been calculated to contain 1000 km³ of salt: the concentration of this great quantity probably took place in a combination of the two methods suggested.

Carbonaceous Rocks: Coal

Coal was the energy source which supported the first 100 years of the industrial revolution. Since 1900 oil has been replacing it to an increasing extent as a source of energy for motive power and heating. Coal is still in demand in the chemical and steel industries, and the main problem of the industry is to discover resources which are economically worth mining. There has been a great increase in open-cast working since 1950, since this requires much less investment in terms of shaft construction and extractive equipment. There is sufficient coal in the rocks of the Earth to last for several hundred years at the present rate of exploitation.

Coal is also interesting as an example of a rock which contains conclusive and detailed evidence concerning its conditions of formation. There are several lines of approach to the evidence provided.

(1) The **chemical composition** of coal varies considerably. The most important constituent, carbon, makes up 60 per cent of peat and over 90 per cent of anthracite (Figure 8.15). Other constituents include oxygen, hydrogen and some nitrogen, and are known as the volatiles, and there may also be variable proportions of materials such as sulphur and mud. The coals with the higher carbon contents have the greater heating values, and are known as coals of higher **rank**.

(2) The **physical characteristic** of coal varies according to the internal make-up: coal may be shiny or dull; dirty and dusty or smooth and clean; it may break into rectangular blocks or may have a more

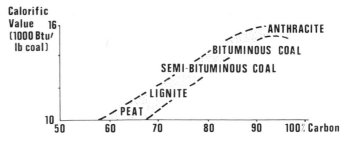

Figure 8.15 Ranks of coal. Based on carbon content and their heat-producing capability.

irregular fracture. All these are determined by the microscopic physical constituents known as macerals (cf the minerals composing the other rocks), which are related to the plant matter from which they originated. The most common macerals are vitrinite (shiny, rich in oxygen and possibly formed from woody matter decaying in water), resinite and sporinite (hard, dull material rich in hydrogen, spores and tars) and fusinite (powdery charcoal, probably formed from the burning of wood). The proportion of each affects the appearance and properties of the coal: a high proportion of resinite and vitrinite with a little fusinite gives a shiny coal excellent for heating since it produces little smoke; more sporinite makes a good gas coal. The combinations of these macerals give rise to **coal types**, several of which may occur together in one seam. The soft, bright coal, which is very brittle and easily breaks up into almost rectangular blocks, is vitrain if it is like structureless black glass, and clarain if it is finely layered. Duller coal is durain, largely composed of sporinite and resinite. Larger blocks of coal are often bounded by dusty fusain.

(3) **Coal occurs in layers, or seams**, like most sedimentary rocks. These occasionally reach over 3 m in thickness, but are mostly less than 2 m, and may be mined if less than 1 m if the quality is good. If a seam is traced very far in one direction it may split into two thinner

seams separated by sandstone or shale, or it may get thicker. In places seams are interrupted by sandstones occupying a channel through the coal: these are termed wash-outs and are thought to be the courses of old rivers (Figure 8.16).

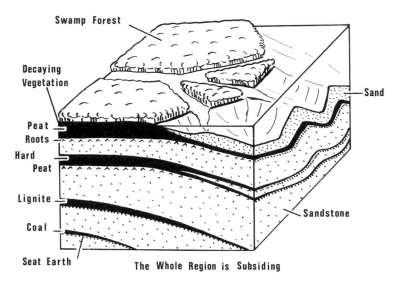

Figure 8.16 How coal is formed. A flat delta is covered by dense, rapidly-growing forests and swampy cut-off lakes. Notice how the peat layers become thinner and split up towards the sea, and the marine sands become thinner as they pass inland. How does deeper burial affect the peat layers? Where will the thickest peat accumulate?

(4) Coal seams are associated with certain types of other rocks. They are often underlain by a seat-earth (the original soil, from which the nutrients have been removed by plants) which may contain the remains of roots: ganisters are sand-grade; fireclays are finer. A shale, or even lime-rich rock, containing marine fossils often lies on top of the coal seam: the fossils in this band may vary from marine animals to fresh-water creatures in their environmental indication. On top of the shale the sediment gets progressively, or suddenly, coarser, and thus largely unfossiliferous, and passes into another seat-earth and coal seam. This sequence (Figure 8.17) is not always repeated in the same form: any

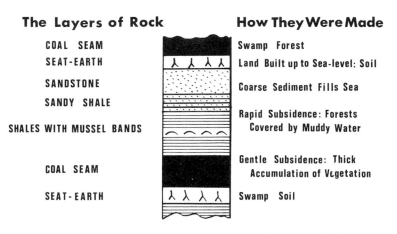

Figure 8.17 Coal seams and associated rocks. This general scheme, with slight variations, is repeated again and again in the Coal Measure series. Each unit is known as a cyclothem.

unit including the coal seam may be missing. Thus in the East Pennine coalfield of England the lowest group of coal-bearing rocks contains thick, escarpment-forming sandstones between the coal seams; above these are a series of more closely spaced coal seams with only shales between and a decreasing number of horizons containing marine fossils; and at the top there are almost unbroken sands with no coals or marine bands. Coals may occur where there is no evidence of any such succession, and in these circumstances they may contain unusual fossils like fish teeth.

(5) Seams of bituminous coal occur in rocks over a wide **range of geological time**, but only those of Carboniferous age are worked extensively on an economic basis in this country. Nearly all the British coal seams are found in the Coal Measure rocks of Upper Carboniferous age. The cyclothem successions are repeated again and again, and in South Wales the total thickness of rocks formed at this time is over 3000 m. The oldest worked seams are of Lower Carboniferous age in central Scotland, and the only late Carboniferous seams occur in South Wales and east Kent: it seems that there was a southward movement of the coal-forming conditions.

There is also an isolated case of a 1 m coal seam of Jurassic age being worked in northern Scotland, and thinner seams occur in the Middle Jurassic rocks of Yorkshire, but the coal-forming processes were not sufficiently developed at this time to give rise to thicker seams. Most Mesozoic and Cainozoic coals are of the lignite (brown coal) variety.

We can now begin to draw together these various threads of evidence and interpret the way in which coal is formed. The carbon content, plant-like elements and underlying root bed show that coal is formed from massive accumulations of land plants. Plants and trees are largely formed of the carbohydrate plant tissues like cellulose (50 per cent carbon; 44 per cent oxygen; 6 per cent hydrogen) plus various fats, resins, oils and waxes, which are all richer in hydrogen (ie the hydrocarbons). The carbohydrates, which form the main body of the plant tissues, are produced, under the action of the sun's rays in the process known as photosynthesis, from the water absorbed through the plant's roots and the carbon dioxide breathed in from the air through the leaves. The resins and fatty substances require in addition mineral salts from the soil for their synthesis: they often serve a protective function.

When a plant dies in your garden it will soon decay completely owing to the action of bacteria and fungi. Their activity is stopped, or reduced considerably, when the supplies of oxygen are cut down. **Peat** is formed in wet, boggy conditions where a very limited amount of oxygen is admitted, and decomposition is confined to the tissue-binding substance and some fermentation of the sugars, forming a carbon-rich deposit with dark brown liquids and jellies permeating the mass together with various gases which escape to the atmosphere:

$$2C_6H_{10}O_5 \longrightarrow C_8H_{10}O_5 + 2CO_2 + 2CH_4 + H_2O$$

cellulose   residue   carbon   methane
                           dioxide

If dying plants fall into deeper, stagnant waters, where there is scarcely any air, only the least resistant plant matter decomposes. The organic remains are mixed with fine, slimy mud, and the whole deposit is known as **sapropel**, which has a higher percentage of hydrocarbons like bitumen.

Thus, besides the original composition of the plants, the type of coal will depend on the conditions in which it was formed. The largest quantities of peat are found on low-lying, flat swamps, such as occur on the surface of deltas (Figure 8.16), where there are also patches of stagnant water in which sapropel forms. The sapropel deposits are usually covered by peat as the pool fills in, or by sands and mud as the streams crossing the delta change course. The weight of all these sediments building up rapidly on the delta causes the whole area to subside, and this allows the individual layers of peat and sapropel to become thicker. They are continuously building up towards the surface of the swamp, but never reach it. The process is interrupted when there is a greater degree of subsidence and the whole area is drowned to become the site for typical marine deposits. The areas in which plants grow most rapidly, and therefore their debris accumulates most swiftly, are those with plenty of sunshine, high temperatures and heavy rainfall. We thus have to look to the tropical parts of the world to find areas where coal is forming today. Near the Equator there is too much rain, and the plant debris is washed away, but the Ganges delta is composed of alternating layers of thick peat and sand such as we might expect.

Bacterial activity ends with the formation and burial of layers of peat and sapropel, but further changes take place in the sediments as they are buried under an increasing thickness of other deposits. The rising pressure and temperature caused by the weight of the overlying matter leads to compaction of the peat, driving off the water and oxygen compounds. This leads to a further increase in the carbon percentage, and the colour changes to black. The whole series of processes finally leads to a hard, brittle coal.

Most of our coals began as peat, but by the time they can be called true coals few of the original plant structures remain obvious for us to see:

(1) **Brown coals**, or **lignites**, are the first in the series of increasing coalification, with a higher percentage of carbon than peat. They still retain some moisture and many of the original plant structures. They are only mined where the higher ranks of coal are absent, as in East Germany, or where it is cheap and easy to use opencast methods, as in the Cologne area.

(2) **Bituminous coals** are the most widely distributed and certainly the most valuable. House coals, gas coals, coking coals and steam coals are all in this group.

(3) **Anthracite** may have as much as 95 per cent carbon and burns slowly without a flame, but gives off great heat. It is harder than bituminous coal and is often greyish and shiny. Anthracite is usually only found in the deepest parts of coalfields, or in highly compressed regions, and it can be thought of as a metamorphic coal.

The sapropel coals are less common and are mixed with inorganic sedimentary matter. **Cannel coal** is mostly formed of the macerals sporinite and resinite, and has a high proportion of volatile hydrocarbon compounds. An unusual feature of cannel coals is that they sometimes contain the remains of fish and other marine life, which are unknown in ordinary coals and suggest transport before deposition. **Boghead coal** is another similar variety, but with more algal remains, giving an even higher proportion of volatile matter: these deposits often grade into oil shales. Coals of this type may be used as a source of gas or oil.

The pattern of sedimentation associated with coal formation and its repetition over and over again is probably due to a combination of the periodic subsidence of the delta area under the weight of accumulating sediment, and the changing positions of the distributary channels. The coal swamp would be drowned and covered by mud and/or sand until the depth of water was filled in and a new swamp established above sea-level, whilst the distributaries would shift position horizontally, as in a modern delta (Chapter 5).

Another fascinating point concerns the formation of so much coal at a particular phase of geological history: many variables need to coincide for such an event. Land plants are known first from the Upper Silurian rocks, and the first forests existed in the Devonian period, limited to lowlying, swampy areas by the fact that they reproduced by means of spores. The Carboniferous period saw a full development of these early plants. The low-lying deltaic swamps with a tendency to subside beneath increasing sediments were particularly extensive across northern Europe and central North America during the later part of the Carboniferous period, and were supplied with plentiful rock debris by the erosion of surrounding uplifted areas. There was thus a combination of plant development and geographical environment, which produced a situation where the energy produced by the sun and converted into plant matter could be stored in the rocks of the Earth's crust: coal has been well-named as a 'fossil fuel'.

Hydrocarbons in the Rocks: Bitumens and Petroleum

Coal is a deposit which has stored the effects of the sun's energy 250 million years ago and preserved it for use today. Petroleum, whilst not strictly a 'sedimentary rock', was formed in a similar way, and is another vital 'fossil fuel'. One of the great problems facing man at the present moment is that we are using up these fuels at a tremendous rate: it is over 100 000 times the speed at which they are forming today, so there is no hope of enough accumulating to replace those that we are using. There is in fact sufficient petroleum and coal to last us a few hundred years, and more may be discovered, but the increased use of these fuels is giving rise to another problem as more and more carbon dioxide is released into the atmosphere. Carbon dioxide helps to trap heat from the sun in the lower atmosphere, and it has been calculated that the increased carbon dioxide content of the atmosphere has raised the average winter temperature in northern Europe by 2–3°C in the last 100 years. This has not affected us much, but it may be connected with the recent melting of many glaciers. If the process is continued until all the carbonaceous fuels are exhausted the temperature will have risen by 12°C, all the ice-sheets will have melted, and the sea-level will have risen to inundate our ports and seaside resorts.

The term petroleum includes a vast array of different substances known as the **hydrocarbons**, which include natural gas (methane, CH_4) on the one hand, and solid asphalt on the other. The paraffin series (C_nH_{2n+2}) is the most common group found in lighter coloured **crude oils** (mixtures of petroleum substances occurring naturally); the napthene series (C_nH_{2n}) is darker and heavier. These minerals occur in oil shales, where heat is needed to distil them, or in liquid form trapped in porous reservoir rocks.

We can tell a lot about coal and its origin by examining fragments of it, but the crude oils found in the rocks have migrated there, and give few clues as to their origin. These deposits are seldom found near coal, or near areas of volcanic action, and it seems most likely that they were formed from minute marine organisms which decayed over a long period in deep, muddy seas. When the muds were compacted into shales the oil was either trapped (oil shale) in the sediment, or was squeezed out with any water and moved into more porous rocks like sandstones and limestones. The oil separated from the water, and accumulated on top of it because it is less dense; if any gas was present that would form a cap on top of the oil. The oil could be trapped in a number of ways, as shown in Figure 8.18.

Geologists are employed by the oil companies to help them in their **search for new sources of supply**. The first step they take is to find out which rocks are likely to be reservoir rocks, and then they have to

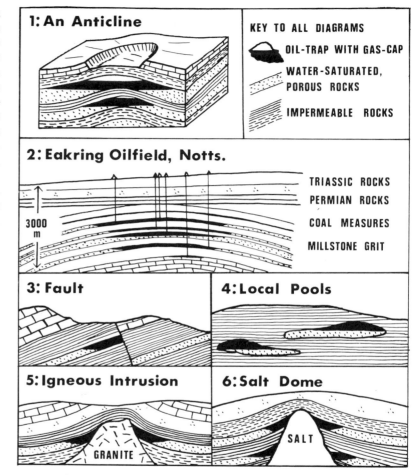

Figure 8.18 Some oil-traps. Various structures involving alternate permeable and impermeable layers help to trap oil, which rises to the top of water-saturated permeable layers.

search for structures affecting these rocks in which the oil may be trapped. This is done by mapping the rocks where they come to the surface, placing particular reliance upon the microfossils which they contain because these can be detected again in the rock chippings brought up by drills. A seepage of oil at the surface will of course give greater confidence in the outcome of such a search, but it may have come from the main deposit some distance underground, and the preliminary survey is still necessary.

If the rocks do not come to the surface, as under the North Sea, geophysical methods of exploration have to be used. These involve setting off explosions to make a small, artificial 'earthquake' and recording the shock-waves as they are reflected from each bedding or boundary plane. A picture of the structure of the subsurface rocks can be built up in this way.

Even when all the geological and geophysical methods of exploration have been tried, there is no guarantee that oil will be present until a borehole has been sunk, and the reservoir has been tapped. This is why microfossils are so important to the oil geologist, since the small fragments brought up by the drill give him a very restricted picture of the rock section.

Britain's resources of oil and natural gas were tiny until 1965. Wells in Nottinghamshire had tapped the Eakring anticline and other small domes since 1940 (Figure 8.18), obtaining oil from Upper Carboniferous sandstone reservoir rocks. More recently small quantities of natural gas have been found near Whitby (Yorkshire) and Cousland (Scotland), but the total contribution of these is very small and the reserves would have lasted only a short time.

One of the world's largest gas fields was discovered in northern Holland in 1959, and intensive surveys of the North Sea area followed. The 1958 Convention of the Seas had allocated mineral rights in the area, and the British government encouraged intensive drilling in its own area by as many concerns as possible. Drilling at sea is very costly, using expensive rigs (£3 million apiece) subject to high insurance rates because of the storm hazards in the area. It costs nearly £$\frac{1}{2}$ million to drill each well—at least ten times the cost on land.

Results in 1965, the first year of drilling, exceeded optimistic forecasts: several productive wells were established off the east coast of England (Figure 8.19) and pipelines laid. Many English homes and

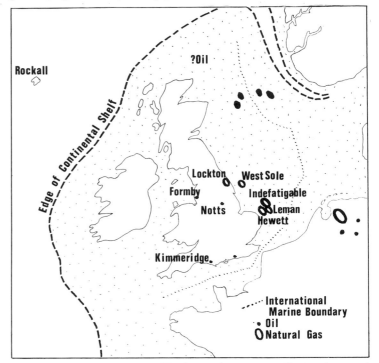

Figure 8.19 British oil and gas. The increasing sources of natural gas and oil in British waters. Further searches are being made in the Irish Sea and in the English Channel.

factories were converted to the use of natural gas by 1970, and supplies for much of the next thirty years assured. The gas comes from a number of anticlines and salt domes (Figure 8.20) which were detected and drilled by newly developed geophysical and offshore

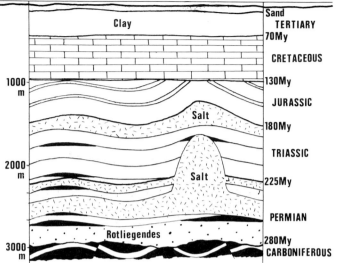

Figure 8.20 The geological situation beneath the North Sea, down to depths of over 3000 m. Which are the chief reservoir rocks and the chief caprocks? How similar is this succession of rocks to that encountered on the British mainland? Note the relatively shallow depth of the North Sea.

extraction techniques. The chief reservoir rock is a porous Permian sandstone, the Rotliegendes, which does not come to the surface on the British mainland, and the gas is trapped by the overlying salt layers. Some gas also occurs in Permian limestone and Triassic sandstones. The gas is unusually pure methane (CH_4), but so far little oil has been found in association. Since 1968 the exploration emphasis has shifted northwards, and large oilfields have been discovered between Scotland and Norway. Other exploration is being carried out in the Irish Sea and to the north-west of the British Isles: watch for newspaper accounts of these developments.

Siliceous, Phosphatic and Aluminous Rocks

The organic, silica-rich rocks are rare. **Diatomite**, composed of millions of the skeletons of the microscopic diatom plants, and **radiolarite**, formed of equally small animals, are both found as soft creamy coloured deposits, often formed in old lakes near volcanoes or geysers. **Siliceous sinter** is built up in bulbous layers round the mouths of volcanic geysers. Nodules of silica occur in many rocks. Where these hard masses break unevenly they are known as **chert**, and this type is common in the Lower Carboniferous limestones of northern England. Some chert layers are now known to be composed largely of radiolaria. **Flint** occurs in more compact nodules, varying from grey to black in colour, having a brittle nature and fracturing conchoidally. These nodules are mostly found in the white chalk rocks of south and eastern England, occurring in lines along the bedding planes, but also across them, indicating that some may have been formed after the main body of the rock (cf Figure 16.7). Many flints have a silicified fossil as a central core. The source of the silica in these lime-rich rocks is a matter for controversy. The silica may have originated as globules which were attracted together into larger masses, often round the shells of dead animals, in a lime-rich sea. Alternatively, it may have come from percolating, silica-rich, groundwater at a much later date.

Phosphate is derived organically from the bones and teeth of animals, or from their excreta. Thus **bone-beds** and **coprolites** (formed of droppings) are common phosphate-rich deposits. Coprolites are forming today off the west coast of Peru, where the guano of bird droppings is many feet thick. **Phosphorite** is the main rock containing phosphate, being a mixture of calcium phosphate and calcium carbonate: it may have been formed in deep, stagnant seas, or by the replacement of limestone as guano-type deposits seeped downwards.

Aluminium is found in all clays, but **bauxite** has been enriched, either by tropical weathering, or by percolating volcanic waters. It is the only rock mined economically for aluminium, and is usually a reddish, mottled deposit of earthy character. The main deposits are in the tropics, but some occur in the USSR, the USA and in southern France (near Les Baux which gave its name to the ore).

Sedimentary Rocks and Time

As layer after layer of sedimentary rocks are piled up on top of each other they preserve a historical record of the different conditions which affected a particular area. We have seen how this works in our interpretation of the Coal Measure rhythmic successions (Figure 8.17). The rocks at the base of a series are the oldest, those at the top the youngest. It is unusual, however, for the same type of rock to extend far in any direction. The oolitic limestones of the Gloucestershire Cotswolds, for instance, pass into deep-water marine clays farther south, and into reefs and deltaic deposits to the north. It is thus difficult to extend our dating of rocks from one part of the country to another based on the recognition of the different rock types alone, since the facies may have changed. We are able to compare sedimentary rocks in different areas because they contain similar fossils. This is where we shall begin the next chapter.

When geologists examine very old rocks, which have been subjected to intense mountain-building movements and overturning, they have to begin by determining which is the top and the bottom of a group of sediments. Such rocks do not contain many fossils, and so sedimentary details, like graded bedding and false bedding, ripple-marks and mud-cracks, are used (cf Figure 12.1). When they have decided which are the oldest and youngest rocks, the geologists are then faced with the even more difficult task of working out the age of the group of rocks as a whole when compared with other rocks some miles distant. Without fossils this was almost impossible before the development of radio-metric dating techniques (described in Chapters 13 and 14).

No one area has a complete succession of sedimentary rocks with the most ancient layer at the base and new layers forming today at the top. There have been interruptions in every part of the world because after a certain period of rock formation earth movements have taken place. The bending and breaking, or folding and faulting of the rocks, plus their uplift, lead to a break in deposition followed by erosion until the landscape of hills and mountains is worn right down and the sea comes in over it. New layers of rock are formed on top of the erosion surface planed across the older tilted or folded rocks. Such fundamental breaks between two phases of deposition are recorded in the rocks as **unconformities**, as illustrated in Figure 8.21.

Figures 8.22 and 8.23 show some of the features of unconformities and successions of sedimentary rocks as they are represented on geological maps. Answering the questions posed under the maps will help you to understand more about the nature of unconformities.

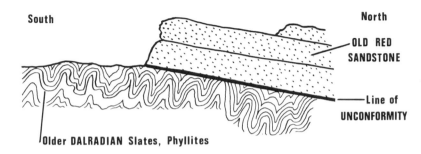

Figure 8.21 An unconformity on the north coast of the Isle of Arran (Loch Ranza). Notice the difference between the folded, altered rocks below the line of unconformity, and the gently-tilted sandstone above. What happened between the formation of the Dalradian and O.R.S. rocks?

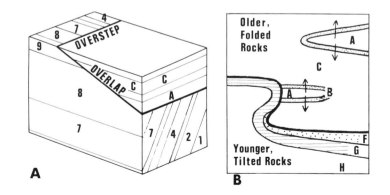

◄ Figure 8.22 Unconformities: overstep and overlap. Use the block diagram (A) and the map situation (B) to devise definitions of overstep and overlap. What is the difference?

An unconformity at Oban, Argyllshire. The conglomerate lies on top of ancient, highly-folded slates and limestones. Write a short account of the history of this area. (*Crown Copyright*)

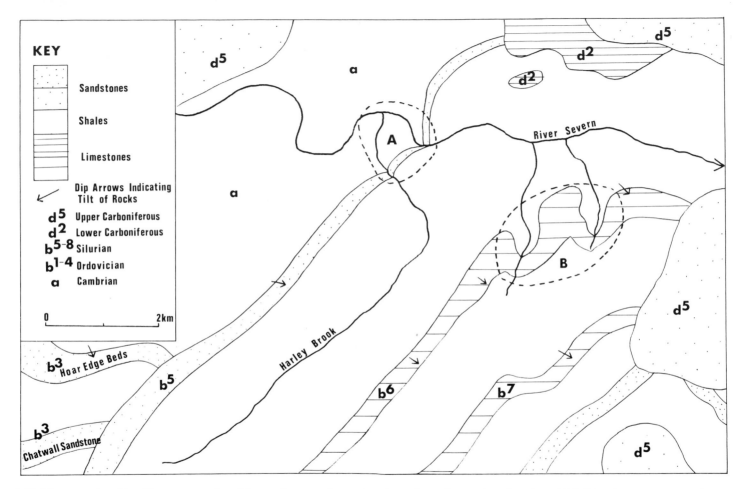

KEY

Sandstones

Shales

Limestones

Dip Arrows Indicating
Tilt of Rocks

d^5 Upper Carboniferous
d^2 Lower Carboniferous
b^{5-8} Silurian
b^{1-4} Ordovician
a Cambrian

0   2km

Figure 8.23 Map exercise. Identify two unconformities, and an overlap and overstep situation. How are the valley outcrops related to the dip directions in the valleys shown ringed at A and B? (Sketch-map based on the Geological Survey 1 : 63 360 Sheet 152, Shrewsbury. *Crown Copyright Reserved*)

THINGS TO DO AND DISCUSS

Investigation 1. Investigating sedimentary rocks

Examine specimens of sedimentary rocks, and make notes in your practical books as follows:

Size of grains:

Colour:

Textures/structures present:

Any fossils?:

Type of rock (a conclusion from the above information):

Locality where found (be accurate): } You can only answer these

Possible age (giving evidence): } if you found the rock yourself

You should be able to identify most of the rocks mentioned in this chapter.

Do not restrict your observations to individual specimens, but examine local quarries and record the associations of rock types (eg alternations of sandstone and shale) and the large-scale features of their arrangement (eg cross-bedding; whether layers maintain an even thickness).

Investigation 2. The uses of sedimentary rocks

You will probably find that sedimentary rocks are being extracted somewhere in your locality: opencast coalpits, quarries in limestone for roadstone or cement, brick pits, sand and gravel pits. Visit these, after obtaining permission from the owners, and note which rocks are being used, the methods of extraction, and the uses to which the rocks are being put. If building work is taking place near you, enquire about the sources of the materials used.

Investigation 3. Classifying sedimentary rocks

We have suggested one method of classifying rocks at the beginning of this chapter. Here is the basis of another, based on the probable conditions of formation. Complete the chart. Which of the two classifications is better for (a) recognising a rock when it is found in the field; (b) interpreting the significance of rock types?

Chart:

ENVIRONMENT	ROCK-TYPE
RIVER	
DELTA	
ESTUARY	
COAST	
SHELF SEA	
etc	

Investigation 4. Geological maps

Read Chapter 20 to help you understand some of the techniques of map reading, and answer the questions set on the maps, Figure 8.3 and Figure 8.23.

Topics for discussion and essay writing

(a) Contrast the modes of origin of limestone and sandstone (O).

(b) Describe briefly the nature and origin of four of the following:

(i) dune-bedding, (ii) loess, (iii) boulder clay, (iv) oolitic limestone, (v) graded bedding (N).

(c) Describe the modes of occurrence and formation of coal. What kinds of rock are usually associated with coal? (L).

(d) Write a short essay on the importance of fossils as formers of sedimentary rocks. (L).

(e) Write a short essay on either (a) rhythmic sedimentation, or (b) evaporite deposits. (O).

(f) Explain the meaning of the term unconformity. Describe an actual unconformity and mention any special features associated with the plane of unconformity (S).

(g) Give an account of the composition and texture of conglomerate, sandstone and shale. Quote examples of these rocks that are found in Britain and indicate the conditions under which sediments are accumulating at the present day (C).

9 Fossils

The Earth, as we saw in Chapter 2, is the only planet within the Solar System on which life can exist in a wide diversity of forms. That diversity becomes apparent as soon as we start to make a detailed study of the living organisms within a few miles of our homes. Even those who live in towns must be aware of the range of living creatures from human beings, through the domestic animals, birds and fishes, to the insects and disease-causing microscopic forms of life. There is also a wealth of plant life, ranging from the tallest trees down to the shrubs, grasses, mosses and microscopic algal forms.

We normally make a major distinction between the animals and plants, because of their fundamental differences in feeding and growth mechanisms. Whilst plants draw inorganic matter from the soil, and convert it into organic substances by reaction with carbon dioxide and water in the presence of sunlight, animals are unable to gain nourishment from inorganic matter and have to feed upon plants or other animals. Such distinctions tend to break down amongst the microscopic, single-celled creatures, and many biologists would place them in a separate group, the protists.

When we extend our studies from our own neighbourhood to the

	PHYLUM	Approx number of species now	Important fossil groups (class or order)	Geological occurrence
ANIMAL KINGDOM	PROTOZOA	30 000	Foraminifera Radiolaria	Ordovician–Recent Cambrian–Recent
	PARAZOA (PORIFERA) (sponges)	4 200		Cambrian–Recent
	COELENTERATA (CNIDARIA)	9 600	Anthozoa (corals)	Ordovician–Recent
	(Several phyla of segmented worms)	27 000	(fossil record scarce)	
	BRACHIOPODA	260	(important fossils)	Cambrian–Recent
	BRYOZOA (POLYZOA)	4 000		Ordovician–Recent
	MOLLUSCA	100 000	Gastropoda Bivalvia (Lamellibranchia) Cephalopoda	Cambrian–Recent Cambrian–Recent Ordovician–Recent
	ANNELIDA (segmented worms)	7 000		Precambrian–Recent
	ARTHROPODA	765 000	Trilobita Crustacea Insecta	Cambrian–Permian Cambrian–Recent Devonian–Recent

entire world, the number of different varieties, or species, of organisms multiplies. A **species** can be defined most simply as a potentially interbreeding group of organisms which is able to reproduce a further generation capable of carrying on this process (thus horses and donkeys, two distinct species, can breed, but their offspring, the mule, cannot). Over a million animal species and a third of a million plant species have been described as existing on the Earth. Our knowledge of the life on our planet is still far from perfect: over 10 000 new species of animals are added to the list each year, and scientists have intimated that the total number of species on the earth may lie somewhere between 2 million and 4 million.

Two questions strike the person faced with a study of this tremendous diversity: how can it be reduced to manageable units for investigation? and what is the explanation for the existence of so many varieties of life on the earth?

Throughout the history of the systematic observation of nature men have attempted to classify what they have seen into groups. The differences between the animal, vegetable and mineral worlds were noted by some of the most ancient writers, but our modern system for classifying living creatures dates from the mid-eighteenth century and a Swedish botanist, Linnaeus. He suggested a more natural type of classification than had hitherto been in use: species showing close similarities could be grouped together as a **genus**. Names were Latinised to give them a world-wide application, and each species is

Figure 9.1 The major groups of animals and plants.

	PHYLUM		Approx number of species now	Important fossil groups (class or order)	Geological occurrence
ANIMAL KINGDOM	ECHINODERMATA		5 700	Crinoidea Echinoidea	Cambrian–Recent Cambrian–Recent
	CHORDATA		45 000	Hemichordata (Graptolites) Vertebrata (fish, amphibia, reptilia, mammals, birds)	Cambrian–Carboniferous Ordovician–Recent
PLANT KINGDOM	CYANOPHYTA EUGLENOPHYTA CHLOROPHYTA PYRROPHYTA PHAEOPHYTA RHODOPHYTA	ALGAE	20 000	Diatoms	Precambrian–Recent
	SCHOZOMYCOPHYTA MYXOMYCOPHYTA EUMYCOPHYTA	FUNGI	48 000		
	BRYOPHYTA		25 000		
	TRACHEOPHYTA (land plants)		250 000	Psilopsida Lycopsida Spenopsida Pteropsida (ferns, conifers, flowering plants)	Devonian–Recent

now referred to by its generic (with a capital letter) and specific names. Thus modern man (a potentially interbreeding group which includes every variety of mankind on the earth today) is known as *Homo sapiens*. He is related by the features of his bodily anatomy to other animals. There are ancient forms of man which are also included in the genus *Homo*. Several genera of manlike creatures form the **family** Hominidae. Together with the families of monkeys, apes, lemurs and tarsiers the Hominidae are members of the **order** Primates. Another step in the hierarchy joins the primates to the lions, elephants, cattle, kangaroos and rodents in the **class** Mammalia. The mammals and

An irregular echinoid preserved in a nodule of flint from the Chalk. Notice the degree of detail preserved (eg plate and pore structures). The scale line here, and in all other photographs, measures 1 cm.

other groups of animals with backbones (fishes, frogs and reptiles) are included in the **phylum** Chordata, and eventually linked with all the other animal phyla in one **kingdom**. As soon as a new species is discovered and named it is slotted into this system, a method having the great virtue of placing it amongst its near relatives. Figure 9.1 lists the major phyla of the plant and animal kingdoms.

But why are all these species represented? How did they come into existence? You can arrive at a part of the answer by examining the habits of a particular animal more closely. It is limited to a certain area in which certain conditions operate; it will eat a restricted variety of foods and may well confine its eating habits to the members of a single animal or plant species. You will notice that it is very well suited to living in such conditions: it is adapted to its particular mode of life in that environment, and the make-up of its body will be such that it is able to move and capture its food supply in the most efficient manner. Study different types of creature in one area, and they will each be found to be specialised for a particular mode of life: birds, for instance, will go for worms or insects or nuts and will have beaks suited to the consumption of their diet. In widely separated parts of the world the same type of environment may be occupied by different groups of animals. Thus in Australia we find that there are a variety of marsupials (feeding their young in pouches), which not only take over the ways of life enjoyed by the more advanced mammals in the rest of the world, but often look like them as well.

Thus one part of the answer to our question is found in the fact that the bodily features and modes of life of organisms are closely related to the different conditions of temperature, soil, water salinity and light in the places where they live.

For the full answer, so far as biologists understand it today, we need to investigate what life was like in the past. What evidence do we have? How reliable is it? To what uses can it be put? This is our study in the present chapter. The animal- and plant-like markings on rocks are the remains of creatures which were once alive. As our knowledge of the past has been extended it has been suggested that only 1 per cent of the total number of species which have lived on the earth are alive today.

It may well have been less than this figure, but in any event there is a vast realm of life in the past to be studied in addition to the one living today.

The Evidence for Life in the Past

A fossil is any piece of evidence in the rocks which tells us something about ancient life. Fossils include the unaltered remains of animals and plants, shell and leaf impressions, and even burrows and footprints. The study of fossils is called **Palaeontology** ('the study of ancient life'). Leonardo da Vinci (1452–1519) was amongst the earliest to recognise the true nature of fossil markings in the rocks:

When the floods from rivers turbid with fine mud deposited this mud over the creatures which live under the water near sea shores, these animals remained pressed into the mud, and being under a great mass of this mud, had to die for lack of the animals on which they used to feed. And in the course of time, the sea sank, and the mud, being drained of the salt water, was eventually turned into stone. And the valves of such molluscs, their soft parts having already been consumed, were filled with the mud; and as the surrounding mud became petrified, the mud which was within the shells, in contact with the former through their apertures, also became turned into stone. And so all the tests (shells) of such molluscs remained between the two stones, that is, the stone in which they were and that which covered them; and these are still found in many places. And nearly all the molluscs petrified in the stones of the mountains still have their natural shell, especially those which were old enough, which would be preserved by their hardness; and the young ones, already calcined, were penetrated in great part by the viscous and petrifying liquid.

Fossils include a wide variety of remains and marks in the rocks. Like living forms today they range in size from the skeletons of enormous animals to a wealth of microscopic forms. Figure 9.2 summarises the ways in which they occur. What strikes you about the parts of animals and plants which are most commonly preserved?

DECREASING USEFULNESS				
NO ALTERATION	ORGANIC COMPOUNDS	SOFT PARTS	Frozen, mummified	EXTREMELY RARE
		SOME HARD PARTS	eg chitin beetle wings	VERY RARE
	INORGANIC COMPOUNDS	MOST HARD PARTS	Calcium carbonate most common in shells; bones of calcium phosphate; noncrystalline silica	ONLY IN RECENT ROCKS
ALTERATION DURING FOSSILISATION	ORGANIC COMPOUNDS	SOFT PARTS	Carbon films, eg plants, worms	UNUSUAL
		HARD PARTS	Carbonised, eg graptolites	COMMON ONLY IN THIS GROUP
	INORGANIC COMPOUNDS	HARD PARTS	Shells pitted; impregnated with minerals in pores; recrystallisation; replacement of original shell material	VERY COMMON
MOULDS AND CASTS	ORGANIC COMPOUNDS	SOFT, HARD PARTS	Imprints in fine-grained sediment, eg bird feathers	UNUSUAL
	INORGANIC COMPOUNDS		Internal or external moulds after shell solution. Cavity refilled as cast	COMMON IN OLDER ROCKS
TRACE FOSSILS			Tracks and trails—signs of movement. Burrows and borings—homes. Coprolites, castings. Tooth-marks	QUITE COMMON

Figure 9.2 Fossilisation. This chart summarises the ways in which animal and plant remains may be fossilised. Notice the relationship between the most common modes of fossilisation, and the degree of information supplied.

The fossils we find in the rocks have undergone a long series of processes since the death of the animal or plant which they represent, and each of these processes removes some of the evidence for life in the past. When an organism dies it will decay, assisted by bacteria and atmospheric weathering, unless it is rapidly removed from their influence by burial in sediment. Soft parts disappear most quickly, but even hard skeletal features are soon lost unless they are buried. When the remains of a dead organism are covered by sediment in this fashion they then become part of the sediment and are subjected to the processes of rock formation which we studied at the beginning of Chapter 8. Compaction may crush the skeleton if it is fragile (Figure 8.1); circulating groundwaters commonly cause the skeletal material to be dissolved away, to be replaced by another substance with some loss of detail, or they may cause the pore spaces in the original shell to be filled by the precipitation of chemical salts. Earth movements, incorporating folding, faulting and even alteration of the rocks by heat and pressure, add to the likelihood of distortion or the complete removal of the fossil evidence from the rocks. The longer the fossil is in the rock, the longer this combination of processes will have to act, and the less chance we shall have of examining the record of life for a particular point in geological time.

We shall thus expect to find fossil remains in sedimentary rocks, and we shall expect to find fewer in older rocks or in rocks formed on the land, where atmospheric oxygen and predators soon destroy any remains.

Look at Figure 9.3 and try to answer the question in the caption. There are certain animal groups which have an impressive and detailed fossil record: we have a good idea of their past history. On the other hand some of the most important and varied groups alive today have hardly any fossil record: the most obvious examples are the insects and worms. The reason for this is straightforward when we understand something of the processes of fossilisation. Animals which have hard (often calcified) internal or external skeletons, bones or shells, are more likely to leave a record of their existence. If you study biology as well as geology you will find that the animal groups you investigate are

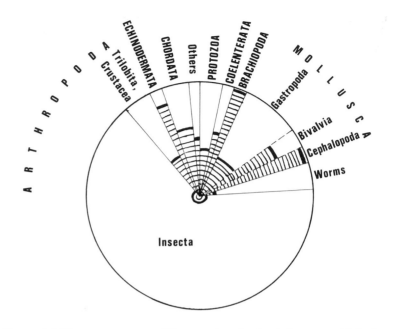

Figure 9.3 There are over a million species of animals known to biologists. This diagram shows how the proportions are grouped into the major phyla. The shaded areas of each zone represent the fossil species known. What does this tell you about the differences between the animals you are most likely to see today and those you are most likely to find as fossils in the rocks?

quite different in the two courses. Some of the most important groups of fossils are extinct (trilobites, graptolites, ammonites, dinosaurs), or almost extinct (brachiopods). Another consequence is that whilst the biologist bases his classifications on the soft parts of animals and plants these are rarely preserved in the fossil state, and so classifications used by palaeontologists are often necessarily quite distinct.

There is thus a built-in bias to the fossil record, and we should always be aware of this. When we examine a group of fossils extracted from a layer of rock we must realise that there is not likely to be any record of the worms, insects or plants which lived in the area of sedimentation represented by that rock layer. We must also be aware of this 'missing factor' when we devise theories based on the fossil record.

We shall now take a close look at the most important groups of fossils you are likely to find in the rocks. The following list relates them to the major animal and plant groups (Figure 9.1) and assesses the main source of fossil remains.

Animals

1 **Corals**: forms within the phylum Coelenterata which secrete calcareous skeletons.

2 **Brachiopods**: a phylum of almost extinct bivalves (ie they have two shells), mostly calcareous.

3 **Lamellibranchs** or **Bivalves** (bivalved molluscs eg clams, oysters, mussels, razor shells)

4 **Gastropods** (eg snails)

5 **Cephalopods**: extinct, shelled forms (ammonites, belemnites)

> the three main classes in the phylum Mollusca: calcareous shelled forms are preserved.

6 **Trilobites**: extinct group in the phylum Arthropoda, having a jointed, calcified skeleton.

7 **Crinoids**: sea lilies, with skeleton of calcareous plates supported on stem.

Echinoids: sea urchins—mobile animals with skeleton of calcareous plates.

> two major groups within the phylum Echinodermata having rigid skeletons.

8 **Graptolites**: an extinct colonial group of uncertain affinities in terms of classification.

9 **Vertebrates**: fishes, amphibians, reptiles, birds, mammals—larger creatures with strong internal skeletons.

Plants

10 Leaf and stem remains, usually carbonised.

This list leaves out a large proportion of today's living creatures. Many of these have a fossil record, but it is sparse, or not obvious to the observer without special equipment. Thus the vast realm of microscopic life gives rise to plentiful microfossils, which are becoming more and more important to geologists in such specialised fields as the search for oil, but they require advanced extraction techniques. Many groups of animals which are less numerous, such as the sponges, bryozoans and other groups of molluscs and echinoderms, also secrete skeletons and are often fossilised, but make up a tiny proportion of the fossil record. There are also the soft-bodied animals which are scarcely represented at all in the fossil record: coelenterates, eg the hydra and jellyfish, many phyla of segmented and unsegmented worms, and most of the huge groups of insects and crustacea with their uncalcified skeletons, as well the greatest part of the plant kingdom. Once again we have emphasised the bias of the fossil record.

THE MAJOR FOSSIL GROUPS

1 The Corals (class Anthozoa of phylum Coelenterata)

Corals are the simplest animals included in our list of the commonly found fossils. Coelenterate cells are arranged in two layers separated by a jelly-like substance (mesogloea), but cell specialisation and organ development are very restricted. Each animal has a large central cavity (the coelenteron) which acts as a stomach, and has a central mouth opening, encircled by tentacles bearing stinging cells. In corals the coelenteron is partitioned by fleshy walls (mesenteries) which increase the digestive powers of the animal by enlarging the area of digestive cells. Some coelenterates, eg the sea anemone, are just soft-bodied, but many corals secrete skeletons. The geologist is particularly interested in these. There are approximately 6000 Anthozoan species today (two-thirds of all coelenterates), plus a further 6000 species of fossil corals.

At the present day there are two distinct groups of skeleton-forming corals, and these are so different that they are placed in subclasses. The **Octocorallia** have eight tentacles and mesenteries and always occur in colonies. They secrete internal skeletons, formed of unconnected calcareous spicules or horny gorgonin: when the animal dies such a skeleton soon disintegrates. These skeletons are often colourful and are used ornamentally. Their structures are simple, including the calcareous tubes and transverse plates of the organ-pipe coral, and the fused spicules of the sea fans.

On the other hand the stony corals produce more massive and resistant calcareous skeletons. These belong to the subclass **Zoantharia** and have similar bodies to the sea anemone with six pairs of complete mesenteries and shorter incomplete fleshy walls in multiples of six between (this group is sometimes known as the hexacorals). The stony corals, or **scleractinians**, secrete septa formed of calcium carbonate crystals between the mesenteries, and these give rise to a pattern of radiating walls. The whole skeleton has the shape of an elongated cup (corallite) (Figure 9.4). Both colonial and solitary forms occur: the solitary forms are often larger (with a corallite diameter of

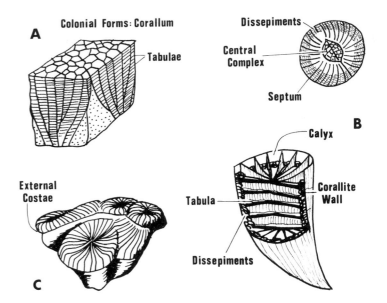

Figure 9.4 The corals. The features of tabulate (A), rugose (B) and scleractinian corals (C) are drawn. Relate these to specimens which you have drawn yourself. What are the differences between the main groups?

up to 250 mm), but the majority are colonial, where the smaller individuals average only 1–3 mm in diameter.

The extensive **coral reefs** (cf Figure 6.17 and Figure 6.18) of the present day are largely formed of the stony corals, but include some octocorals. The colonial stony corals live in close symbiosis (ie living together for mutual benefit) with algae, and so are limited to shallow, well-lit waters where the algae can photosynthesise. They only flourish in tropical areas with temperatures of between 25°C and 29°C and need copious oxygen. These reefs are associations of a remarkable variety of creatures and are amongst the richest living communities in this sense on the earth, but their existence is threatened by the plague of 'Crown of thorns' starfish, which is stripping the stony coral areas of the Australian Great Barrier Reef and the many Pacific and Indian Ocean islands. The delicate balance of the coral reef food chain has been upset, possibly by the human demand for attractive *Triton* shells: this large marine snail eats the starfish and previously restricted its effect. We wait to see whether the starfish plague will cause the stony corals to become extinct, or whether this is part of the predator–prey cycle whereby the predator will increase in numbers until its food supply is almost wiped out; the predators then starve to death and the prey has an opportunity to re-establish itself. Our answer should not be long delayed, since already 25 per cent of the Great Barrier Reef coral and up to 90 per cent around some Pacific islands is destroyed.

The **solitary corals** are not limited in depth by an algal association: they occur over a much wider climatic zone and in water of low temperatures down to depths of 7000 m.

Fossil Corals. The skeletons of corals, like other fossils, are filled with mud or mineral matter after death; they may be bored through and through by sponges and bivalves; and in older rocks the whole mass may be recrystallised, destroying the organically secreted structures.

The stony corals (**scleractinians**) are the only modern, skeleton-forming corals which are important in the fossil record: they form the reefs and solitary fossil corals which occur in the Mesozoic and Cainozoic rocks (Figure 9.5). The **octocorals** have few fossil ancestors.

Two other groups of fossil corals are found in the Palaeozoic rocks: both became extinct before the development of the modern corals. The **tabulate** corals are the first to occur in the geological record and were important in the late Ordovician, Silurian and Devonian periods, but were later overshadowed by the **rugose** corals. The tabulate skeleton is simpler (Figure 9.4), with few internal structures, and these corals were always colonial. The rugose skeletons are more complex, with a variety of internal features more akin to the scleractinian group: they are common in rocks of middle Silurian, Middle Devonian and Lower

Coral: *Lonsdaleia*, Lower Carboniferous. Note how the corallites are joined together: the smaller examples are juveniles growing in the spreading colony. Draw a portion of this rugose coral, labelling the parts shown.

TERTIARY			Octocorals (few fossils)
CRETACEOUS			
JURASSIC		Scleractinians	
TRIASSIC			
PERMIAN	EXTINCT	EXTINCT	
CARBONIFER'S		Rugose	
DEVONIAN			
SILURIAN	Tabulate		
ORDOVICIAN	First Coral Fossils		
CAMBRIAN			

Figure 9.5 The geological history of the corals.

117

Carboniferous age, but then gradually became less important and died out. Many of these rugose corals are used as zone fossils: this means that they are used in dating the rocks in which they occur (the principle is further explained in the section 'Fossils and time' at the end of this chapter).

Reef corals of these distant periods may have demanded similar conditions to the reef corals today, and if this is assumed we can obtain an indication of the ancient environment in which they were formed. Fossil reefs, however, have a different distribution from those today: this suggests that either tropical conditions were more widespread in the past, or the land masses must have moved their positions relative to the climatic belts, or that the distribution of reefs is controlled by special conditions necessary for the symbiotic algae, and that the reef corals have become increasingly dependent on these with time. The solitary coral fossils may indicate deeper water conditions for sedimentation of the rocks in which they are found.

2 The Brachiopods (phylum Brachiopoda)

Brachiopods form a small, rare group today: they are sometimes called the 'lamp-shells' because the shape of the shells of one group resembles the old oil lamp. The most distinctive feature of the brachiopod body is its food-catching organ, the lophophore (Figure 9.6). This is a fold in the body wall which is covered in fine hairs, or cilia: as these move they direct a current of water towards the mouth where microscopic food particles can be extracted. The animal is enclosed between two shells, or valves, but most of the space is occupied by the lophophore and by the complex system of muscles used for opening and closing the valves. The bodily organisation is more complex than the coelenterates, but is still at a low level: there are few specialised sense organs, digestion is by means of cells in the gut wall, the blood circulation is very weak and there are no special gas exchange organs (ie gills).

Externally the brachiopod shell is similar in some ways to an important group of living molluscs, the lamellibranchs: palaeontologists must learn to distinguish the features of the two groups. Biolo-

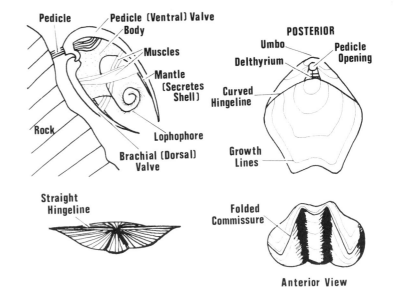

Figure 9.6 The brachiopods. The main features of living brachiopods and their shells. Identify similar features on the brachiopod fossils you find.

gists did not separate these two bivalve groups until the nineteenth century. (The features of the two groups are compared in the next section.)

The brachiopods living today are the remnant of a much more important past: 260 living species compare with 30 000 fossil species. *Lingula* burrows in the sands or muds of warm, shallow seas close to the shore and can survive less salty estuarine conditions ('brackish' water). Its valves are not hinged, but are held together and adjusted by complex muscles. Similar shells are found as fossils in rocks since the Cambrian system (ie 600 million years old), showing little change in shape and generally occurring in sediments interpreted as being of coastal origin. All other living brachiopods are also bottom-living (benthonic) crea-

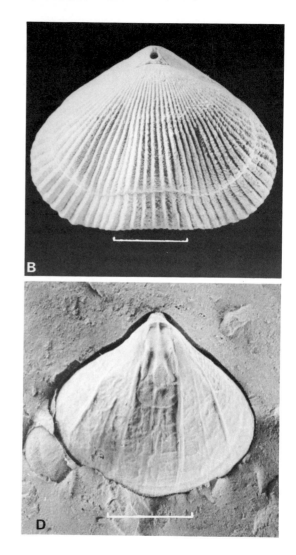

Some brachiopods. A and B show the features of the outside of the shells: what are the main differences between the two? C has been distorted as the rock in which it is entombed has been affected by earth movements. D shows a cast of the internal features of the shell (the shell itself has been dissolved away), including muscle scars and even blood vessels.

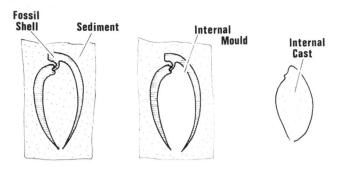

Figure 9.7 Moulds and casts. Write your own definition of the difference between a cast and a mould.

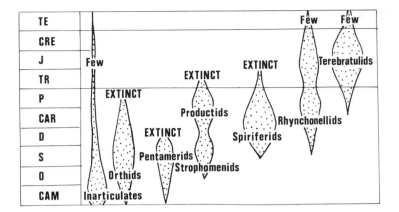

Figure 9.8 The geological history of the brachiopods. What age would the rocks be if you found productids, spiriferids, rhynchonellids and the occasional terebratulid in them?

tures. Forms like *Terebratula* attach themselves to the rocky sea-floor by a fleshy pedicle and have a sedentary existence; their shells are hinged and vary from shorter, heavier and more rounded shapes in shallower and rougher water to more fragile, elongated forms in quieter conditions. All brachiopods are marine, but hardly any live beyond the edge of the continental shelf (ie 100–200 m deep).

Fossil Brachiopods. The Palaeozoic brachiopod fossils do not normally have the shell preserved as such, but only the moulds and casts of the internal or external features (Figure 9.7). Mesozoic shells have either been replaced or recrystallised, and only the most recent fossils retain the original shell materials. This illustrates the effect of the passage of time on the preservation of fossils.

Palaeozoic brachiopods are particularly important in the fossil record (Figure 9.8). The earliest known forms, the Cambrian, did not have hinged shells and include *Lingulae* like those we find today: the group without hinged shells is known as the **inarticulates**. Hinged (**articulate**) brachiopods became really important in the Ordovician and Silurian: genera like *Orthis* with a long, straight hingeline, and the pentamerid *Conchidium* with its bulbous shape and rounded hingeline,

were particularly common. Spiriferids, with a straight hingeline at the widest part of the shell, were important in the Devonian, and the Lower Carboniferous rocks contain many productids. At the end of the Palaeozoic most of these groups died out, leaving the small, ribbed rhynchonellids, which tended to repeat certain varieties of shell form at intervals in the fossil record, and the smoother-shelled terebratulids: these two groups were at their most important during the Mesozoic, since when there has been a further decline to the present situation.

It is probable that brachiopods have always inhabited shallow seas, and therefore they can be used as a criterion for such an environment. Many of them existed over a very short time range, being rapidly replaced by newly developing forms. Some brachiopods are thus useful as zone fossils, and in spite of their sedentary adult habit the larval forms are free-swimming and ensure a widespread distribution of the group. Brachiopods are most important in this way in the lime-rich rocks of the Devonian, Lower Carboniferous, Middle Jurassic and Upper Cretaceous.

3 The Molluscs (phylum Mollusca)

The molluscs include a wide range of animals, and the phylum is the second largest today, including over 150 000 living species. Oysters, snails, squids and the octopus are common members, and many molluscs secrete hard calcareous shells: this means that they are amongst the most numerous forms in the fossil record, since it seems that the molluscs have always been a large and successful group. Common fossils, eg the ammonites and belemnites, are extinct molluscs.

Even the simplest molluscs show an advance in bodily development compared with the corals and brachiopods. All molluscs with the exception of the Bivalvia have a mouth region equipped with a scraper-like tongue (the radula), respiration is by means of complex gills or even lungs, the blood circulation and excretory system become well-organised, and movement in the search for food is more efficient. The most advanced molluscs are amongst the most complex and highly organised of all invertebrate animals: squids, for instance, can swim as fast as fishes, digest their food as rapidly as advanced mammals, and their nervous system enables them to change their body colours over a wide array within seconds.

Bivalve Molluscs: Lamellibranchs or Pelecypods. This group of filter- and deposit-feeders is the most sedentary of the three major mollusc classes: few move far, and a great number are fixed to the rocks or pebbles beneath the water in which they live, or plough into the soft sediment to varying depths. They secrete bivalve shells round their bodies, and these act as a support and protection. Food enters the shell, is sorted on the gills and swept into the mouth. The animal only extends a muscular foot outside the shell to pull itself through the sediment, or, if it is burrowing, a pipe-like pair of siphons to connect it with the clear water and the supply of food above the sediment. Those forms which are cemented to the surface, like the oysters and mussels, have hardly any foot and no siphons. No lamellibranch possesses a head or radula: the group is highly specialised for its particular mode of life and feeding.

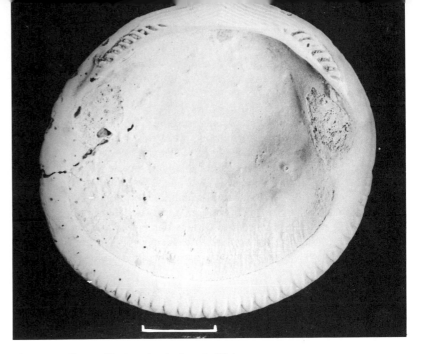

Bivalve mollusc: *Glycimeris glycimeris*, Pleistocene.

Fossil Record. The shells of lamellibranchs are common Mesozoic and Cainozoic fossils, although preservation is less common in older rocks, as is the case with other shell creatures: once again the oldest records are merely impressions of the mould or cast variety. Lamellibranch shells are composed of an outer dark organic layer (never fossilised), and a calcareous, layered shell consisting of varying proportions of aragonite and/or calcite: most fossilised forms show a high proportion of the calcite. The features of their shells are shown in Figure 9.9.

121

Lamellibranchs are important fossils throughout the geological record from the Cambrian period (Figure 9.14). The Palaeozoic forms show an increasing range of diversity. At first they were mostly attached in some way, but by the Carboniferous period there were burrowing and swimming varieties, as well as freshwater forms. The most advanced forms, which became more active burrowers, developed through the Mesozoic and into the Cainozoic, and the thick-shelled, sedentary oysters also appeared at this stage. There is a correlation between the thicker-shelled varieties and shallow, warm, lime-rich seas, in contrast to the thinner shells associated with deeper or freshwater conditions. The lamellibranchs outnumbered the brachiopods by the Mesozoic era and became the dominant bivalves. It is probable that the competition between the two groups was settled in favour of the more advanced and better-adapted lamellibranchs.

Lamellibranchs are particularly characteristic of shallow waters and stand a very good chance of fossilisation if their mode of life involves burrowing into the sea-floor sediment. They may also be used as zone fossils in circumstances where other groups are absent. Thus the non-marine lamellibranchs are used for this purpose in the Coal Measures of the Upper Carboniferous; the lowest Jurassic rocks in South Wales are divided according to the Gryphea sequence (Figure 9.10); and the British lower Tertiary rocks are partly zoned by the lamellibranchs because of the great variety and numbers contained.

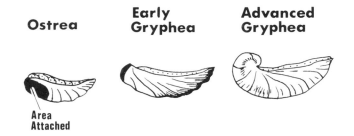

Figure 9.10 Gryphea evolution. This series of stages from an oyster to the coiled Gryphea took place on several occasions. It is possible that the excessive coiling led to the extinction of each line, but the same trend was repeated again later.

Lamellibranchs and brachiopods are similar in many ways, including their modes of life and the shapes of their shells. It is thus important to be able to distinguish between them, and there are several lines along which this is possible (Figure 9.11).

(1) In life the lamellibranch valves are to the left and right of the hinge, which faces upwards in the active varieties, whilst the shells open downwards (ventral); sedentary forms, like oysters, rest on their right valve. The brachiopod valves are normally one on top of the other (ie the pedicle, or ventral, valve on top of the brachial, or dorsal, valve).
(2) The two brachiopod valves can each be divided into two equal halves (ie they are equilateral), but the lamellibranch valves cannot (inequilateral). A little care is necessary, since some lamellibranchs are almost equilateral.

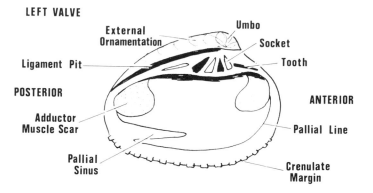

Figure 9.9 The bivalve mollusc. You can often see the internal features of these shells, since they fall apart on the death of the animal. External features such as radiating or concentric ribbing, and the shape of the shells are also distinctive.

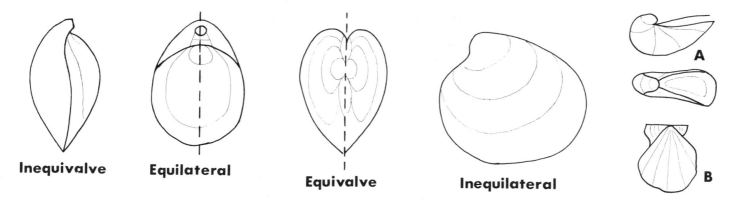

Inequivalve   **Equilateral**   **Equivalve**   **Inequilateral**   A   B

Figure 9.11 Brachiopod or bivalve mollusc? What do the terms used here mean? Which apply to the brachiopod, and which to the bivalve? In which category would you place the examples (A) and (B)? Why are these two slightly different from the normal rule?

(3) Lamellibranchs normally have two identical valves (ie they are equivalved), but the two brachiopod valves are always of different sizes (inequivalved), the pedicle valve being the larger. Some of the attached lamellibranchs (oysters) have unequal valves, but they are also obviously inequilateral. It is common, however, for the valves of lamellibranchs to become separated on death, whereas the brachiopod shells are more commonly held together.

Thus, whilst there is some variation within the forms of lamellibranch shells, the brachiopods are always equilateral and inequivalved.

(4) Brachiopods often have a prominent pedicle opening, which may be in the form of a round opening at the top of the pedicle valve, or a triangular gap. Lamellibranchs do not have this feature.

(5) If the hingeing mechanism can be seen, lamellibranchs will have mirror image patterns of teeth and sockets in each valve, whilst brachiopods have only two teeth in the pedicle valve, and two sockets in the brachial valve. The separation of lamellibranch valves on the death of the animal means that their internal features are seen more often in fossils than those of brachiopods.

(6) The interiors of lamellibranch shells have the scars of only one or two muscle attachments, and the same pattern occurs in each valve. Brachiopod shells have more muscle scars, and different patterns in each valve.

(7) Lamellibranch shells are opened by a ligament, housed in a pit outside, or inside the hingeline. Brachiopods do not possess such pits.

(8) Another interior feature, unique to the lamellibranchs, is the pallial line, which marks the outer edge of the mantle attachment to the shell. When siphons are developed for deeper burrowing there is a deep notch in this line, known as the pallial sinus.

4 The Gastropods (class Gastropoda, phylum Mollusca)

This mollusc class includes snails, slugs, limpets and several varieties of tiny swimming or planktonic organisms. Being molluscs they have many points in common with the lamellibranch bodily organisation (eg a muscular foot used for movement, gills, blood circulation and nervous system), but in addition they have a mouth area with radula and a head in which the sense organs are concentrated, they come out of their shells for moving around and are found in a vastly greater range of environments. The most distinctive feature, however, concerns the arrangement of the body. It seems that, although a primitive, shelled mollusc may have had a head and mouth at one end and a cavity containing the gills and excretory organs at the other, the gastropod larva went through a process known as torsion which brought the rear cavity to a position above the head in the adult gastropod. Gastropod shells, apart from specialised shapes like the conical limpet, are coiled, and at first this took the form of a flat spiral, only found in the early, and now extinct, bellerophontids. As the animal expanded the space available in its shell, it adopted a helicoid spiral form (Figure 9.13): this necessitated a redistribution of weight, and the shell assumed a position with its axis slanted with respect to that of the body. As a result one side of the animal's bodily develop-

Gastropod: *Clavella longaeva*, Eocene.

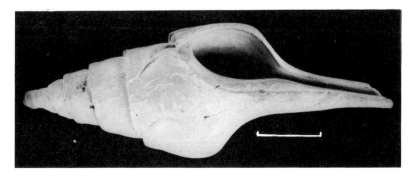

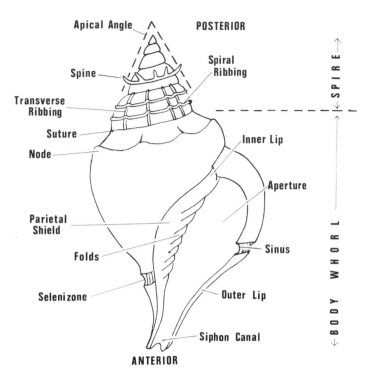

Figure 9.12 The gastropods. The general features of gastropod shells: no single animal ever had all of these on its shell!

ment was restricted, and paired organs like the gills have become reduced to one.

The marine gastropods include the shelled bottom-crawlers, which feed in various ways: some scrape rock-encrusting algae from the sea-floor; some scavenge through sediment and extract organic matter; others are carnivorous and reach their prey by boring holes into protective shells or by direct attack. In addition many gastropods have

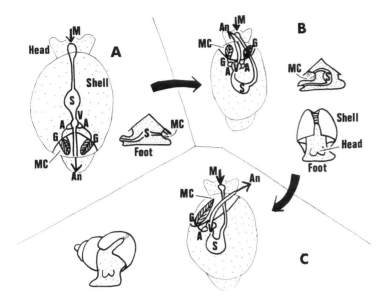

Figure 9.13 Gastropod development. The development of the gastropods from an ancestral form (A), for which there is no evidence, to a typical Palaeozoic form (B) with a plane coiled shell and a modern form (C). A process known as torsion took place between (A) and (B), in which the back end was twisted round and above the front. A later development led to the spiral shell, and the progressive suppression of one set of any paired features (eg gills). A—auricle; V—ventricle; M—mouth; S—stomach; G—gill; An—anus; MC—mantle cavity.

lost their shell, or have reduced it so that swimming and floating are possible. On the land there are freshwater snails with thinner shells, and land snails and slugs, which have modified their gill chamber into lung-like organs for breathing air: all these land varieties feed on plant matter.

Amongst this immense variety of forms, adapted to a great range of environments, the shelled gastropods are of most interest to the palaeontologist. The shapes of gastropod shells do not vary greatly through geological time and are in fact often repeated. One of the most distinctive features is the presence of a canal at the anterior margin of the aperture in some forms (Figure 9.12): this protects an inhalant siphon which projects forward to sample the water ahead of the animal and separates it from the exhalant organ, and is a feature of the more advanced and carnivorous varieties.

Fossil Record. Gastropods are more plentiful today than they have ever been (Figure 9.14), and there are more varieties of gastropod than of any other mollusc group. Palaeozoic and Mesozoic forms are not common, and it is only since the early Cainozoic that they have really become important. In such rocks they are sometimes used as zone fossils and are particularly valuable in distinguishing between marine (thick, well-ornamented shells sometimes having siphon canals) and freshwater (thin, simple shells) conditions in the formation of sediments.

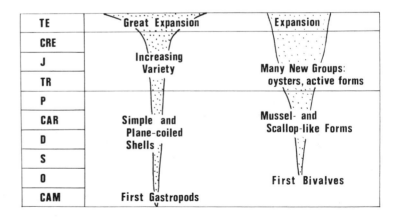

Figure 9.14 The geological history of the gastropods and bivalve molluscs. Compare the history of the bivalves with that of the brachiopods.

125

5 The Cephalopods (class Cephalopoda, phylum Mollusca)

The squids, octopuses and their close relations are the most highly developed animals without backbones. Their degree of muscular co-ordination and sensory development is only rivalled by the higher vertebrates. They are quick-moving, carnivorous predators, and although the octopuses are not quite as formidable as they are portrayed in science fiction (the largest has a body 30 cm long), some of the largest squids are up to 16 m in length, including the tentacles.

These creatures have, however, lost any hard shell they may have possessed, and they have a poor fossil record. And yet there are over 10 000 fossil cephalopods to compare with the 400 existing species. Fossil cephalopods are related most closely to the three species of *Nautilus* living in the Indian and Pacific Oceans. These have coiled shells, like the extinct ammonites and their close relations, and their bodies show some marked differences from the squids: *Nautilus* tentacles are short, it has four gills as opposed to two, and the funnel which it uses as a form of jet engine is formed in a different way. The animal lives only in a part of the coiled shell, the rest being divided into chambers (camera) by walls of calcium carbonate (septa). The chambers are filled with gas, which gives the animal buoyancy in the water. The *Nautilus*, however, is slow-moving and clumsy compared with the squid.

Fossil Record. The fossil record is thus particularly important in this group. The **nautiloids** are the earliest to appear and originally had straight shells: once they became coiled tightly their shells showed little record of further changes (Figure 9.16). The **ammonoids**, which are nearly all coiled shells in a plane spiral (Figure 9.15), are found commonly in the rocks of Upper Palaeozoic age, where they are known as **goniatites**, and in the Mesozoic rocks, where they are known as **ammonites**: these separate names reflect two phases of development, each of which gave rise to a rapidly-changing variety of forms and ended in a major wave of extinctions (Figure 9.16).

The cigar-shaped *belemnites* (Figure 9.15) were important fossils only during the Mesozoic, although first reported in Carboniferous

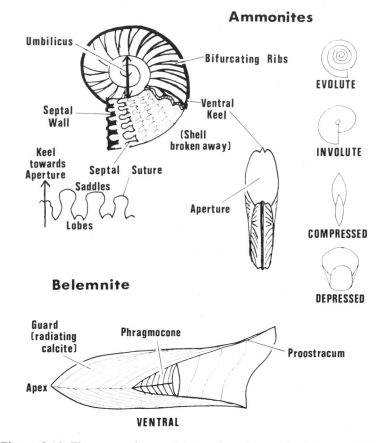

Figure 9.15 The ammonites and belemnites. The main features of these fossils: neither have living representatives.

Cephalopod: ammonite, *Dactylioceras commune*, Lower Jurassic.

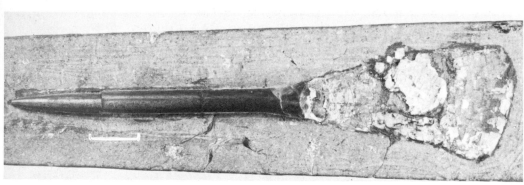

Cephalopod: belemnite. The guard is preserved together with part of the proostracum.

rocks, and they became extinct with the ammonites in the late Cretaceous. The belemnites are assumed to represent the internal structures of the ancestors of today's squids, although these animals have now lost all but the faintest remnant of their internal skeletons.

Goniatites, ammonites and belemnites are all used as zone fossils, and the first two are the most important of all groups in this respect. Many of them must have been extremely mobile and are widely distributed geographically. New forms rapidly succeed old in the rock succession, and they are extremely common fossils. The Jurassic rocks in particular can be divided into very narrow time zones on the basis of the ammonites they contain.

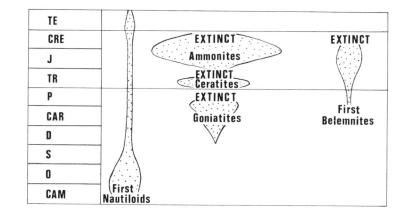

Figure 9.16 The geological history of the cephalopods.

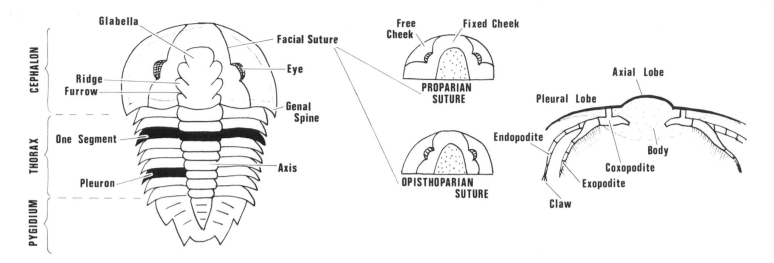

Figure 9.17 Trilobites. The main features of these fossils are shown on the left, and details are enlarged on the right.

6 The Trilobites (class Trilobita of phylum Arthropoda)

This is another group of fossils with no living representatives, like the ammonites, belemnites, ancient corals and most brachiopods. The segmented nature of the fossilised skeleton (Figure 9.17), together with the fact that some trilobite fossils indicate that this was an exoskeleton, shed several times as the animal grew, have led to the classification of the trilobites in the vast phylum **Arthropoda**, which today includes three-quarters of a million species of insects, spiders and crustacea (crabs, lobsters). Trilobites were the most primitive of these arthropods and are amongst the oldest animals found with a plentiful fossil record: they occur in the earliest Cambrian rocks. Like the other arthropods the trilobites possessed jointed legs and advanced sense organs, such as the oldest known examples of compound eyes (similar to those in bees). Most crawled on the sea-floor, but some probably burrowed in the sediment (streamlined shape, no eyes), floated or swam through the water (large area of spinose skeleton, eyes beneath body), or lived in darker deep zones (large eyes).

Fossil Record. All our knowledge of the trilobites is derived from the fossil record, although it is important for us to attempt to establish their relationships with modern animals.

The first trilobites in the Cambrian rocks (Figure 9.18) are the oldest zone fossils. At this stage they are characterised by forms with

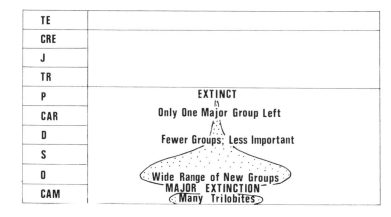

TE	
CRE	
J	
TR	
P	EXTINCT
CAR	Only One Major Group Left
D	Fewer Groups; Less Important
S	
O	Wide Range of New Groups
CAM	MAJOR EXTINCTION / Many Trilobites

Figure 9.18 The geological history of the trilobites.

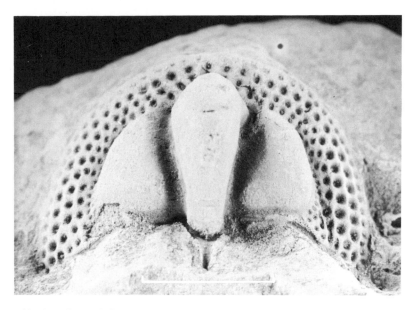

spinose skeletons, small tail-shields, and if eyes are present they are attached to the glabella on the headshield. Some of the largest trilobites, up to 45 cm long, occur in Cambrian rocks. The following Ordovician period witnessed an almost complete change and replacement by new families with a much wider adaptation to different marine environments. Silurian forms were similar, but at the end of that period many groups became extinct. The decline continued through the Upper Palaeozoic, as the fishes and ammonoids provided increasing competition, and final extinction came in the Permian period (about 225 million years ago).

The trilobites thus provide us with an almost complete record of the progress of a major group of animals: we miss only the earliest stages of their history (which is true of virtually every major group), and we can study their increasing importance, the height of their development and their decline to extinction.

Trilobite: *Cryptolithus concentricus*, Upper Ordovician. Compare this headshield with Figure 9.17. Which features differ from the 'typical' form depicted there? It is common to find trilobite headshields, or tailshields, separated from the rest of the skeleton.

7 The Echinoderms (phylum Echinodermata)

This name means 'spiny-skinned' and is a good description of such members of the group as sea urchins and starfish. It is not so accurate for others such as the sea lilies and soft sea cucumbers. The essential features of the members of this phylum are that they are all marine creatures living largely inside a shell, or test, of interlocking calcite plates (or separated calcite spicules) and have a unique internal circulatory system known as the water vascular system which is particu-larly concerned with respiration and locomotion. In many forms the plate arrangements can be divided into five zones (Figure 9.20), but in others the symmetry of the test is essentially bilateral. To effect growth of the test, new plates are secreted on the margin of the apical system, enlarged as they reach the widest part, and trimmed down near the mouth.

A variety of forms inhabit different marine environments. Some have thicker and stronger shells and live close to the low tide mark, often burrowing into sand or making a home beneath a rock; others live in deeper water, moving across the surface of the sediment in search of prey; and the sea lilies are often attached to the sea-floor by stems of varying lengths, using a system of arms to direct the rain of fine organic debris from the upper layers of the ocean into a central mouth.

Fossil Record. The possession of a hard, calcite-plate skeleton means that many groups of echinoderms are common fossils: the original shell material is commonly permeated by additional chemical matter and made considerably heavier. In addition, the many forms in which living tissue divided the calcite plates tended to disintegrate on death, leaving piles of calcitic detritus. Many of the Carboniferous limestones of Britain are composed largely of such fragments, sorted by marine currents before final deposition.

The greatest variety of the fixed, stem-based sea lilies (**crinoids**, Figure 9.19), together with several extinct related varieties of echinoderm (eg **cystoids**, **blastoids**) is found in Palaeozoic rocks (Figure

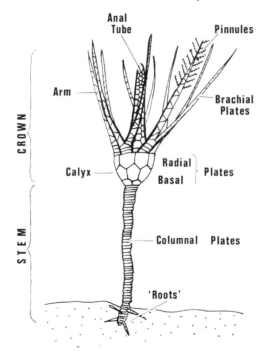

Figure 9.19 A typical crinoid. These animals have skeletons composed of interlocking plates of calcite. Fossils of anything apart from the columnal plates are rare.

Figure 9.20 Echinoids. The features of a regular (A) and irregular echinoid (B). Compare the two sets of features and relate them to the modes of life of the animals.

130

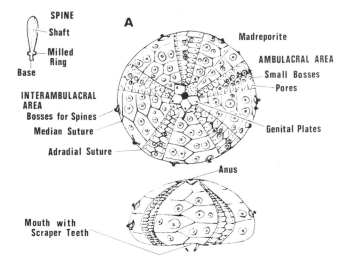

SPINE
Shaft
Milled Ring
Base

A

INTERAMBULACRAL AREA
Bosses for Spines
Median Suture
Adradial Suture

Madreporite

AMBULACRAL AREA
Small Bosses
Pores

Genital Plates

Anus

Mouth with Scraper Teeth

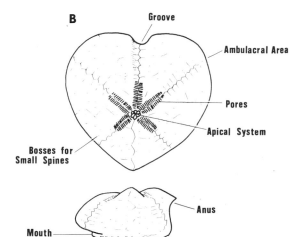

B

Groove

Ambulacral Area

Pores

Apical System

Bosses for Small Spines

Anus

Mouth

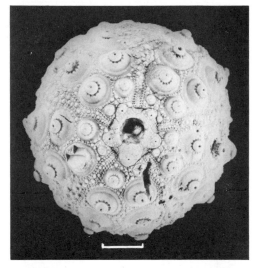

Echinoid: *Hemicidaris*, Jurassic. Compare this form with the *Micraster* form.

Echinoid: *Micraster*, Upper Cretaceous.

131

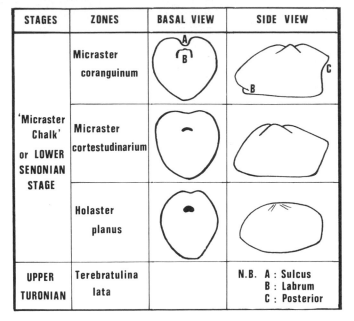

STAGES	ZONES	BASAL VIEW	SIDE VIEW
'Micraster Chalk' or LOWER SENONIAN STAGE	Micraster coranguinum		
	Micraster cortestudinarium		
	Holaster planus		
UPPER TURONIAN	Terebratulina lata		N.B. A : Sulcus B : Labrum C : Posterior

9.22), but complete specimens are rare, and it is most common to find the separated calcite plates. Only one group of crinoids survived into the Mesozoic and Cainozoic, where stems and cups, including those of free-swimming, stemless varieties, are found in the Middle Jurassic and Upper Cretaceous rocks.

Sea urchins (**echinoids**) are rare as fossils until the Mesozoic, where they are commonly associated with shallow, clear, lime-rich seas including reefs. They are particularly common in the Upper Cretaceous Chalk, where the evolutionary sequence of Micraster has been suggested (Figure 9.21) and the group is used for zoning. Sea urchins seldom live deeper than 200 m and their presence in the Chalk shows that the rock was formed in relatively shallow water.

◀ **Figure 9.21 Evolution in echinoids.** Notice the changes in particular features as time advances (towards the top of the diagram).

Figure 9.22 The geological history of the crinoids and echinoids.

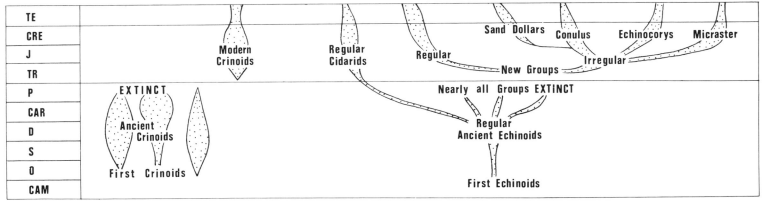

8 The Graptolites

The graptolites are an extinct group of tiny colonial animals, which leave markings on ancient shales and silts resembling the blades used in fretsaws (Figure 9.23). They have been extinct for 300 million years and bear such little resemblance to any living creatures that it has been difficult to associate them with any major group of animals. The colonial form at first suggested that they should be assigned to the coelenterates with the corals and many other small colonial creatures. Modern opinion tends to place them closer to the very primitive members of the phylum Chordata because of the distinctive way in which they build their skeletons: this phylum includes forms which have no skeletal backbone to support the interconnecting notochord, as well as the higher vertebrate animals.

Graptolites: *Didymograptus*, Ordovician (left).
Climacograptus, Upper Ordovician.

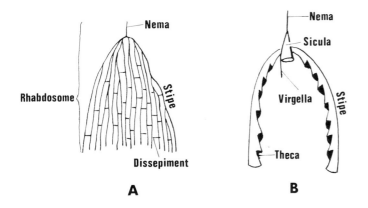

Figure 9.23 Graptolite features. Simple diagrams showing the many-branched dendroid form (A) and the more typical graptoloid form (B).

Fossil Record. Graptolites form extremely delicate skeletons and their fossils are largely restricted to fine-grained sediments deposited in quiet conditions; some are preserved in silts or sands, but with less detail.

The lowest Ordovician rocks contain many-branched (**dendroid**) forms, which were probably attached to the sea-floor by a stem. The branches were soon reduced in numbers, and the two-branched, 'tuning-fork' varieties became important. Further modifications appeared rapidly in the Ordovician rock succession, involving the alteration of the orientation of the branches until they coalesced into a single, double-sided branch (Figure 9.24), and the elaboration of the individual cups in which the animals lived. By the Silurian the graptolites were nearly all single-branched with the cups (thecae) on one side only; a second period of elaboration was experienced with lobate,

133

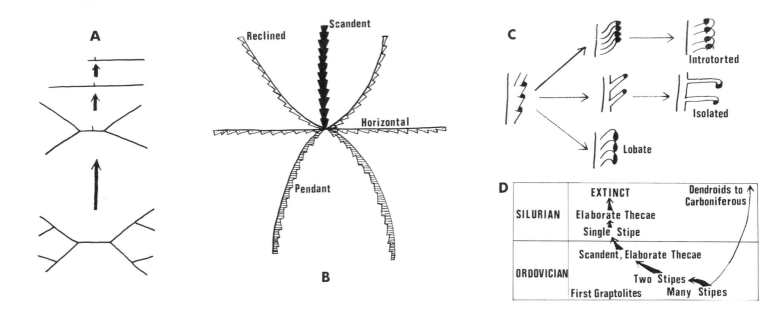

Figure 9.24 Graptolite development. Graptolite development took place in a variety of ways. The simple features were altered by reducing the number of stipes (A), mostly during the early Ordovician; by changing the direction in which the thecae faced (B), a process which took place in the later Ordovician; and by elaborating on the basic form of the thecae (C)—in the late Ordovician scandent forms and in the Silurian. The general progress within the group is shown in (D).

hooked and isolated forms, but all except for a few dendroids became extinct by the end of the Silurian. It is thought that most of the graptolites with a small number of branches lived a floating existence in the plankton, since their fossil remains are found with a similar distribution to other planktonic forms, ie in sediments formed in conditions which were unfavourable to bottom-living organisms so that the only fossils they contain have dropped in.

The rapid changes in skeletal shape and the widespread geographical distribution of the graptolites make them good zone fossils for the fine-grained rocks of Ordovician and Silurian periods. Much of the history and structure of complex areas such as the Southern Uplands of Scotland, and central Wales, has been worked out after a study of the graptolites contained by the rocks. It is a pity that the fragile graptolite skeleton disintegrated when incorporated in coarser rocks, or when involved in the rougher conditions of sedimentation in which only shelly fossils like brachiopods, trilobites and corals were preserved, but occasionally graptolites do occur in such rocks and enable their age sequence to be tied into the finer, graptolite-rich shales.

9 The Vertebrates: from Fishes to Man

All the groups of animals so far mentioned are invertebrates, having no backbone. The vertebrates (fishes, amphibians, reptiles and mammals including man) are a very important group today, and their development is particularly interesting to us as human beings. All the vertebrates are members of the phylum Chordata: their vertebral column supports and protects the main nerve chord.

Fossil Record

The fossil remains are generally much larger than those of the other groups with which we have been dealing, and are less common. The bones of fishes are often extremely delicate and are preserved only in fine-grained sediments. Many vertebrates live on the land, where weathering processes and scavengers soon cause even the most massive bones to disintegrate. The palaeontologist may discover a huge fossil 'graveyard' with the bones of hundreds of animals, but is often restricted to a piece of bone from a limb, skull or vertebral column, or even just a few teeth. The sharks, for instance, have a non-calcified skeleton which is not preserved, and their history is virtually known only from their fossilised teeth.

A study of the different groups of vertebrate fossils illustrates their development (Figure 9.25).

(1) The **Agnatha** are the most primitive fish-like animals, and have no jaws or paired fins. They are the oldest fossil vertebrates: scales occur in the Ordovician rocks of Wyoming and their remains become more common in the Upper Silurian and Devonian rocks. These early varieties were mostly ostracoderms, which had bony plate armour over their head regions and lived until the end of the Devonian in mainly freshwater conditions. The lamprey is a living representative which has adopted a parasitic mode of life.

(2) The **placoderms** were another early group of fish-like creatures, but had primitive jaws and paired fins. They contained a variety of forms: some, like the ostracoderms, were small (10–30 cm) bottom-living forms, but others were highly mobile predators, like *Dinichthys*, which was 10 m long. They were particularly important in the Devonian period, but became extinct before the end of the Permian.

(3) Another group, of which we have a fossil record beginning in the Devonian period, includes the **sharks and rays**. They have a more complex body, including more specialised jaws, and a shape which is well-adapted to their role as fast predators (sharks), or bottom-living (rays). Sharks are known mainly from their teeth fossils, but also from carbonised film impressions on fine-grained rocks. They have altered very little through geological time, surviving many competitors, eg the huge marine reptiles of the Mesozoic and the whales of the Cainozoic. It seems that they were at their largest (15 m long) in the Miocene.

(4) The Devonian period is often known as the 'Age of Fishes', for besides the three groups already mentioned we also find the first fossils of the **true or 'bony' fishes** in rocks of this age. These fishes have strong, calcified skeletons, and have become the most common types: in fact they include more species than all the other groups of vertebrates combined—eels, cod, flat-fish, lung-fish, sturgeon, marine and freshwater types. The bony skeletons are preserved frequently as fossils and sometimes the patterns of scales can be detected.

Reptile vertebrae: Jurassic.

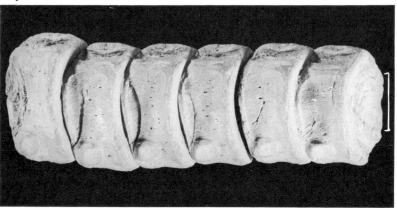

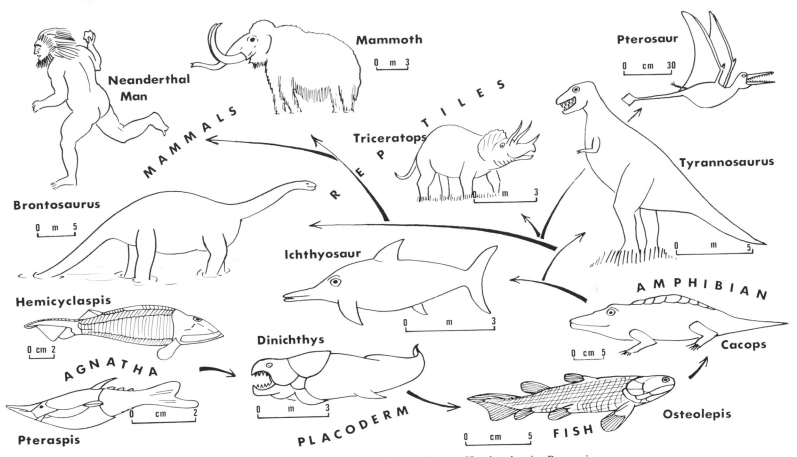

Figure 9.25 From fishes to man. Ostracoderms: Hemicyclaspis, Pteraspis (Silurian, Devonian); Placoderm: Dinichthys (Devonian); Bony Fish: Osteolepis (Devonian); Amphibian: Cacops (Permian); Reptiles: Brontosaurus, Ichthyosaur, Pterosaur (Jurassic); Triceratops, Tyrannosaurus (Cretaceous); Mammals: Mammoth, Neanderthal Man (Pleistocene).

Many of these bony fish have fins with a ray-like pattern of supporting bones (Figure 9.26), but a small group have a lobe-like pattern which gives a much stronger support. It is thought that the latter feature, combined with other skeletal developments and the ability to breathe air instead of water-filtered oxygen, was part of a series of changes leading towards the amphibian group, which first appeared later in the Devonian.

(5) The **amphibians** are the simplest group of vertebrates able to live on dry land, though they have to return to water to lay eggs: larvae grow in the aquatic conditions and breathe through gills, and the process of metamorphosis is necessary for the change to the adult state. The earliest Devonian varieties had short legs and could only drag themselves slowly along the ground, but by the Permian they were more mobile, and their longer legs could lift them clear of the ground. They then had to meet competition from the reptiles, and now only a few groups, such as the frogs and salamanders, are left.

(6) The **reptiles** were the first group of vertebrates able to live completely out of the water. The earliest, Upper Carboniferous, varieties looked very much like the amphibians of the time, but had the advantage of being able to lay eggs with a protective covering. From the Permian to the end of the Cretaceous these animals dominated the scene.

The earliest reptiles were small creatures, and although they were freed from the aquatic life by their reproductive capabilities they were still bound to the water areas by the availability of food. It seems possible that it was not until the Permian that meat supplies in the form of large plant-eating reptiles became available over a wide area of land, a development which led to diversification of the group as a whole. Some of the Permian varieties had characteristics which are more reminiscent of the mammals, such as a false palate separating the mouth and nasal passage, and it is thought that this group, or some close relation, gave rise to the mammals.

Many of these varieties are confined to the Permian period, and they were replaced by a wide spectrum of new groups in the Triassic, including the land-based dinosaurs, gliding and flying pterosaurs, and

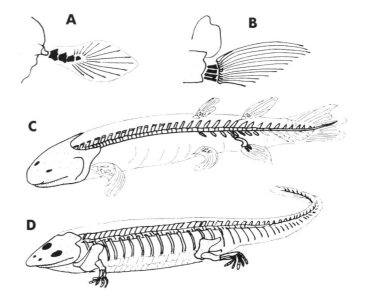

Figure 9.26 From fishes to amphibians. A is the lobe-fin, different from the normal ray-fins (B) of the fishes: this formed the basis for developments towards land vertebrates. C is a crossopterygian fish, which had lobe-fins and could breathe air. It is thought that this may have developed to an amphibian (D). What features of these two animals suggest similarity of bodily form, and what features are related to their mode of life in water (C) and on land (D)?

the ichthyosaurs and plesiosaurs which returned to marine conditions. These included the largest creatures on the Earth until the end of the Cretaceous, when most became extinct. The reason for this is a great puzzle, especially as many of the groups were at the peak of their development. Suggestions for solutions have included a change in climate, or the results of the changes in vegetation during the early Cretaceous period, but none are satisfactory. Today the snakes, crocodiles, turtles and lizards are the remnants of this once all-powerful group.

(7) The **birds** are feathered and warm-blooded flying animals. The earliest bird-like creature is recorded from the Jurassic rocks of southern Germany. It had feathers, but reptile-like teeth, and is thought to be a link between the groups. By the Tertiary there were large numbers of flightless birds like our present-day ostrich.

(8) The **mammals** are the dominant group of animals throughout the world today, but although the first mammal fossils are found in rocks of Triassic age, the group spent over 100 million years as an insignificant minority overshadowed by the reptiles. The wholesale extinction of the reptiles at the end of the Cretaceous (ie just over 70 million years ago) enabled the mammals to spread into many of the environmental 'niches' vacated on land and in the sea. The mammals are warm-blooded, are mostly covered with hair or fur, and have the most advanced and adaptable bodies of all the animals living on the Earth.

The primitive groups include the egg-laying monotremes (eg duck-billed platypus) and the marsupials (eg kangaroo and opossum) which protect their immature young in a pouch. Ninety-five per cent of the mammals alive today are of the third, placental group, named after the organ which supplies nourishment to the embryo in the womb: many of these animals are born almost ready to face the rigours of the world around. The placental group includes many hoofed mammals (eg odd-toed horse and rhinoceros; more plentiful even-toed cattle, pig, deer, camel, elephant), the edentates with smaller teeth (eg armadillo, sloth), the rodents, the flesh-eating group (eg dog, cat, bear) and the primates (eg lemur, monkey, ape, man). Many of these have well-documented histories in spite of the fact that fossil preservation is unusual on the land where scavengers and the agents of decay are at their busiest (Figure 9.27).

The history of the development of the vertebrates is particularly interesting because it provides us with a record of the diversification and replacement of a wide range of animal types through the geological record. Whereas nearly all the major groups of invertebrate

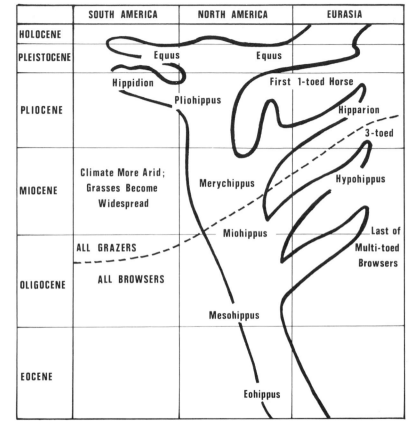

Figure 9.27 Evolution of the horse. Eohippus was half the size of a Shetland Pony, and had an arched back, four hoofed toes and low-crowned teeth (for browsing on leaves). Mesohippus was more horse-like, though only 60 cm high; it was three-toed, had a larger brain and there were sharp crests on its teeth. Merychippus was up to 1 m high at the shoulder, and developed a deeper jaw to hold the resistant, high-crowned teeth suitable for chewing grasses.

creatures appear almost suddenly in the Cambrian or Ordovician without any record of their earlier connections, a more adequate family tree of the vertebrates can be constructed. Many of the groups of vertebrates can be linked by fossil skeletons showing intermediate characteristics, but we shall probably never be able to answer many of the questions concerning some of the more dramatic steps involved. How did jaws develop in the primitive fishes? Why did some forms move out on the land? How did the reptiles begin to lay protected eggs? When and how did the mammals and birds acquire warm blood together with methods for regulating their body temperatures?

10 Plants

Fossil plants are less common than animals and although they have played a vital part in the development of life on the Earth, we can reconstruct only a poor and fragmentary picture of their history. Many plants are soft, or susceptible to rapid decomposition when they die, and it is only rarely that carbonised films of leaves and stems are preserved: rapid burial in fine sediments is necessary, and this means that the high degree of compaction associated with muds squashes plant remains flat. Occasionally tree stems may be preserved in sandstones with the original round cross section (Figure 8.1). It is common to find the leaves separated from the stems, and they are often transported before entombment in sediment.

Plants: Fossil tree trunk, Jurassic (left).
        Linopteris, Upper Carboniferous (Coal Measures).

Fossil tree stumps and roots, Glasgow. The rock has been removed to expose these relics of an ancient forest. (*Crown Copyright*)

The **simplest plants** are those without any circulatory system of their own. Most of them live in water or in very damp conditions. They include the single-celled diatoms and other planktonic forms, which may give rise to thick deposits of their siliceous skeletons; the blue-green algae, which often form large colonial structures and are amongst the earliest fossils known in the Precambrian rocks; the parasitic fungi; and the mosses and liverworts. All are rare as fossils, though the first two groups are occasionally important.

Plants living on the dry land need a more advanced circulatory, or vascular, system and complicated reproductive facilities: roots, stems, leaves, spores, flowers and seed have developed. The **earliest vascular plants** (Silurian of Australia, Devonian of Britain) have a simple stem lying along the ground and topped by bare branches carrying spores (Figure 9.28). By the Carboniferous period there were dense forests of giant horsetails, tall spore-bearing lycopod trees, ferns and early seed-bearing conifers. Other groups of seed-bearing plants, such as the ginkgos, became important in the Mesozoic: all these early seed-bearing varieties are known as **gymnosperms** ('bare seed'). Our present flowering plants, or **angiosperms** ('hidden seed'), which make up 95 per cent of all plant species today, only developed in the early Cretaceous, and grasses only became widespread in the Miocene.

This is the general pattern of plant development. It is interesting to note that the first land plants came into being at a similar date to the land animals (amphibians and insects); that the Cretaceous development of flowering plants was accompanied by that of pollinating insects; and that the spread of grasses in the Miocene caused many mammal groups, eg the horse, to change their feeding habits from browsing on leaves to grazing (Figure 9.27).

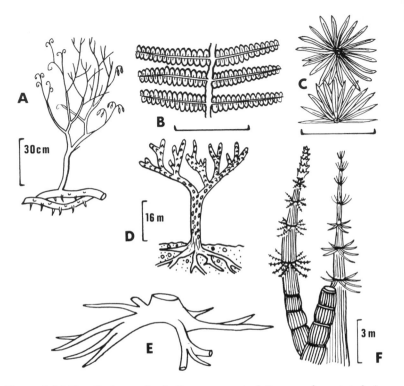

Figure 9.28 Fossil plants. *Scale line represents 2·5 cm, unless stated.* A—Psilophyton (Devonian); B—Pecopteris; C—Annularia; D—Lepidodendron; E—Stigmaria (roots); F—Calamites (B–F all Upper Carboniferous).

11 Microfossils

The fossils we have mentioned so far are those you will find in the rocks you examine with the naked eye. Most of them are between 1 cm and 8 cm in length, although the vertebrate remains in particular may be much larger. None of them, however, are of much use to the palaeontologist who is examining the cores or rock chippings brought up from a depth during drilling. Such a person is dependent on the microfossils, which are present in most rocks in great numbers. Just as there is a wealth of microscopic life on land and in the sea, so there is a vast realm of microscopic fossil evidence. A whole new branch of palaeontology has arisen from the study of these tiny forms, and has contributed important new techniques of investigation. The microfossils can be used in the same way as normal fossils for giving a geological date to the rocks in which they occur and for providing information concerning the development of life on the Earth, or the conditions in which the rocks were laid down.

The size of microfossils makes it inevitable that they will be of use only to specialists, since their study requires the use of sophisticated extraction techniques, the possession of expensive equipment and a knowledge of the latest discoveries in the rapidly enlarging field of identification.

Microfossils are drawn from several of the animal groups, including those from which important macrofossils are drawn.

(1) Protozoans of the order **Foraminifera** are widespread and abundant, especially in the Mesozoic and Cainozoic rocks. They are some of the simplest animals, having no cellular division of their body, although the more advanced types may produce calcium carbonate shells with many interconnected chambers. These tiny fossils are used widely in the search for oil.

(2) **Ostracods** are minute relations of crabs and lobsters (ie Arthropods), living in the water and related to the water-fleas. They have a bivalve shell covering their body, and the wide variety of ornamentation on this shell is amazing for such tiny forms. Different types of ostracod today are very sensitive to changing salinity, so the fossils may be used for interpretations of past salinities.

(3) **Plant spores** are found in many different rocks, including those formed in the sea, as well as those on the land. Coal seams and the overlying shales can be correlated across country on the basis of the spore assemblages they contain.

(4) New groups are coming into use as the study of microfossils gains ground, although it will be many years before they attain a widespread significance. One such group is known as the **Hystrichospheres**, which were once part of the plankton floating at the surface of the oceans; another is the **Acritarchs** of more uncertain affinities. All marine sediments contain these microscopic forms which have spherical, ellipsoidal or polygonal shapes ornamented with ridges or growths (spines or tubes, branching or interconnected). One gram of shale may contain thousands of specimens, which can be studied after removal from the crushed rock. Because they originated from floating plankton they are widely distributed throughout the world: forms of Jurassic age in England and Australia are identical. Many groups show signs of rapid development into new varieties, another important feature to the geologist. When more is known about these microfossils they will surely help us to gain a fuller picture of the history of our planet.

FOSSILS AND TIME

Certain facts emerge from a close study of the fossils contained in the rocks. A pattern can be distinguished in the development of life on the Earth, even when one bears in mind the bias imparted to the record by the processes of fossilisation. A number of aspects of this pattern are particularly significant.

(1) The record of **the earliest life on the Earth** is extremely sparse. There are hardly any fossils in the rocks over a span of 3 000 million years: some organic compounds have been recognised from extremely early rocks (eg Fig Tree Series of South Africa, c 3200 million years old), and there are remains of the most primitive plant life from other rocks of later date (eg Gunflint Chert of Lake Superior region, c 2000 million years old), but no geologist would pretend that the record is as full as he would like. There is no indication of the actual origin of life: it seems that the most primitive forms were the earliest, but the manner and date of their emergence, or creation, are unknown. A number of theories have suggested answers to this problem, involving life emerging from organic compounds present in the special conditions of a primeval sea. We can never know the complete facts concerning the origin of life. Laboratory studies may indicate how life could have, or might have, emerged, but not how it started.

(2) **The numbers and variety of living things have increased** through geological time: compare the situations in the major groups during the Silurian, Carboniferous, Jurassic and Eocene periods, ie at roughly 100 million year intervals. The increase has not been a steady process, but has been punctuated by periods of more spectacular expansion, like the Ordovician and Jurassic, as well as by others which witnessed widespread extinctions (Permian, late Cretaceous). After the very slow Precambrian progress within single-celled creatures, the Cambrian system of rocks presents us with a wide range. This includes not only the first records of metazoan (many-celled) creatures, but the representatives of almost all the major groups of animals. From that time to the present there is a record of increasing complexity, and a diversity which spreads from confinement in the sea to gradual mastery of the land and air.

(3) During the course of this process of expansion of life on the Earth **new species** of animals have developed, and others have become extinct. The arrival of the different groups of vertebrate animals illustrates this, and on a smaller scale fossil groups have been examined closely and show tiny, but definite, changes of shape with the passage of time (Figure 9.21). Gaps in the records are due to the fact that many of the vital 'links' have not been preserved as fossils, and we saw earlier in this chapter that the scales are weighted against the chances of any animal leaving a record of its existence. As the descent of a particular group of animals is traced through the rocks, modifications clearly happen: this process is known as evolution. Some animals have shown rapid evolutionary changes (eg graptolites, ammonites), whilst others have shown little or no change over the whole range of geological time in which fossils are commonly found in the rocks (eg brachiopods such as *Lingula*).

(4) Another concept which is useful in the interpretation of the fossil record is that **the environment** clearly affects the features of an animal's anatomy. This is obvious when we compare the shapes of, for instance, fishes living in the buoyant medium of water and their distinctive methods of locomotion and respiration, with those of the land vertebrates, which have limbs to lift their bodies clear of the ground so that they can move efficiently in the less buoyant medium, air. It is also instructive to compare the shapes of the extinct reptile, the ichthyosaur, and the dolphin, a mammal: they are almost identical, and both have lived in similar conditions.

(5) The increasing numbers of animals and plants on the Earth's surface eventually led to **competition** between the species. Each environment could support only a limited number, and the entry of a new, better-equipped, or adapted, species would lead to the replacement and extinction of the old variety, unless it could readapt itself in another environment. Only one complete phylum has ever become extinct, and that is a rather small group of Cambrian sponge-like creatures, the Archaeocyatha: representatives of all the rest which are first found as

fossils in the Cambrian and Ordovician rocks are still with us. As well as competition the extinction of groups of animals must have been due to changing environmental conditions. If an animal became too specialised a slight change would cause its extinction and this could lead to a whole series of extinctions, since the interrelationships in nature tend to be very complicated. Yet many animals survived drastic changes, and this must have been due to a very adaptable body structure and mode of life.

The present animal and plant populations of the world are the result of 'descent with modification' from primitive ancestors of the main phyla. This concept is well-supported by the evidence we have studied. Within each species there is much variability: just look at the human race, all members of the species *Homo sapiens*, but with so many variations in skin colour, height, hair, nose, etc. Such diversity extends to all species of animals and plants, and in many is fostered by nature's bounty in reproduction: salmon lay as many as 28 million eggs each in a year. Some members of each species must be slightly better adapted than the others to their particular environment, with the result that they will survive more easily and have a greater effect on the next generation in reproduction. The study of genetics has helped to show how vital are the hereditary genes in determining the characteristics of the new populations.

Fossils record the general line of evolutionary progress, although the detailed observations concerning the mechanisms of change have to be made on living animals. We can also derive an idea of the slowness of the processes, since geological time is so vast. It has been calculated that the diameter of the horse's tooth changed as it turned from leaf-browsing to grazing—by 1 mm in 5 million years. And this was one of the more rapid changes!

(6) The development of life through geological time is also important in a very practical way for the geologist: it is the basis of nearly all his **time correlations**. Now that radioactive dating is becoming of greater significance we shall be able to fill in some of the gaps in the historical picture and give more accurate numerical dates, but fossils are the only way of comparing the relative ages of sedimentary rocks in widely separated areas.

When we have studied a group of rocks in one area, and established their age sequence by such methods as those suggested at the ends of Chapters 8, 10 and 11, it is vital that we should be able to compare them with other rocks elsewhere. We can often trace a band of distinctive rock across country: the ridge of Chalk across the centre of the Isle of Wight is an outstanding example. Aerial photographs and geophysical methods help us to do this on a larger scale, but it is dangerous to put complete trust in such methods, for the conditions in which a certain type of rock are formed may migrate across a region with time: the same rock-type may be formed earlier in one place, and later elsewhere.

Fossils provide us with the most reliable method of telling whether two rocks are of the same age. Each rock contains a certain assemblage of fossils, and those which were formed more recently will have more highly developed and modern-looking fossils. This is due to the results of the process of evolution combined with rapid geographical distribution of new species. Each rock, or part of a rock, or group of rocks, with a unique assemblage of fossils is known as a **zone**. Such zones are referred to by an exclusive species of fossil they contain, and are the smallest units of rock-time division. Several zones form a **stage** (cf Figure 9.21). Zones vary greatly in thickness since the rate of rock-deposition has not always varied in the same way as the rate of evolution.

The ideal **zone fossils** would be animals which have had a wide geographical sphere of influence; have evolved rapidly and show rapid changes as they are traced upwards through the rocks; have been able to live in a variety of environments; are of a relatively small size (ie 5–10 cm) so that they can be seen as a whole and easily extracted from the rocks; and are commonly preserved. In practice zone fossils may possess many, if not all, of these qualities, but some rocks have to be zoned by the least likely fossils, since they contain nothing else. The graptolites (Ordovician, Silurian), goniatites (Devonian-Permian), ammonites (Jurassic, Cretaceous) and several groups of microfossils are the best fossils for this purpose. Refer back to your studies of each group and pick out the reasons for this. Free-floating crinoids and

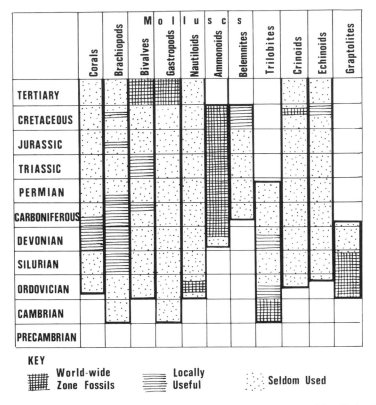

KEY

World-wide Zone Fossils   Locally Useful   Seldom Used

Figure 9.29 Zone fossils. The usefulness of the various groups of fossils in the relative dating of the rocks in which they occur. Which are the most useful? Notice the outstanding contribution of the molluscs.

some trilobites are also good, and where none of these are present corals, brachiopods, belemnites, freshwater mussels, plant spores and fish have been used (Figure 9.29). Why are these groups not such good zone fossils?

The main difficulties arise when a rock changes its facies, and it has often been the case that a different set of conditions was not liked by the animals living at the time of formation. Thus graptolites are typical of the fine-grained, deepwater shales of Ordovician and Silurian age, but are seldom preserved in the coarser, shallow-water deposits of the same age which have different fossils (mainly trilobites and brachiopods). It is usually possible, however, to find a rock layer where the graptolites and trilobites occur together.

There have been notable cases where the study of fossil zones has helped geologists to work out the structures affecting the rocks of an area. This was the case in the Southern Uplands of Scotland, where the study of the graptolite zones led to the discovery that the rocks were tightly folded. Palaeontologists assist in the search for oil by tracing oil-bearing horizons across country and determining the structures affecting the rock successions.

FOSSILS AND PLACE

Fossils may also assist us in attempts to reconstruct the environment in which they lived. Ecology is the science which studies the relationships between organisms and their environment today; palaeoecology is the study of ancient life environments. Whilst the ecologist is sometimes embarrassed by the wealth of data available to him and the complexity of the relationships he observes, the palaeoecologist is often embarrassed because either his data are too few (owing to the poor fossil record), or because there is a mixture of fossils which once lived in the place of sedimentation and those shells which have been transported to the area from elsewhere or those which have dropped into the sediment from the plankton or swimming creatures. The palaeoecologist has to allow for the fact that a large proportion of the living forms have not been preserved as fossils.

Animal and plant shapes are related to their living functions. Thus it is possible to distinguish between lamellibranchs which burrow to different depths by the extent of their pallial sinus; regular sea urchins roam about in any direction on the surface of coral lagoon sediments, whilst irregular echinoids burrow in the sediment; and the features of vertebrate skeletons and jaws reveal whether the owner was a marine or terrestrial beast, how active it was, and what was the main component of its diet. These features are preserved in fossils. The most useful fossils to the palaeoecologist are those which are buried in the place where they lived: sedentary forms like oysters, corals, brachiopods and crinoids, and burrowing forms like many lamellibranchs and some gastropods and echinoids.

Trace fossils are becoming increasingly important, since they are also features formed by animals as the sediment accumulates. They include the tracks (straight-line trails or meandering grazing tracks), burrows, resting marks and subsurface sediment-browsing systems. One snag is that it is seldom possible to recognise the originator of the features, since they leave no remains: many crustacea and annelid worms are responsible for such burrows today. Certain types of trace fossils are associated with particular rock facies, and they may also be used to work out the rates of sedimentation.

Other new techniques have also increased the scope of such studies. The proportions of the different isotopes of oxygen in the calcium carbonate of ocean-floor deposits are related to the temperature at which the shells of the microscopic animals were formed near the ocean surface. The pollens in thick deposits of peat give a very good idea of the trees living there in the past, and of the temperature and humidity of the climate at the time. A picture of the falling temperature at the end of the Tertiary era, for instance, can be obtained by comparing the animal and plant life of the time.

As in the case of the study of zone fossils the study of palaeoecology has an economic application: certain rock facies, identified by their fossil content, are known to be good oil reservoirs. It becomes more and more important to study the complete relationships of the fossils when we find them; not only do we have to recognise the type in order to date the rocks which contain it, but we should note the rock type, the position of the animal in the rock and the group of fossils of which it is a part.

THINGS TO DO AND DISCUSS

Investigation 1. Introduction to fossils

It is important to relate the fossils to living organisms of today and to examine the different ways in which fossil material may be preserved in the rocks.

(*a*) Examine a group of named fossils from the school collection. Which of these are like modern animals or plants?

(*b*) Which modern organisms are not represented?

(*c*) Which parts of an animal are preserved as fossils? Is the whole animal preserved, and how much detail can be discovered from the part that is (eg shape, structural features, colour)?

(*d*) Make notes concerning the evidence you can detect relating to the modification of the organism (or the fragment preserved) from the time of its death up to the time when it was extracted from a rock (eg What has happened to the original organic features?).

Investigation 2. Examining fossils

Answer the following questions for each of the major groups (corals, brachiopods, lamellibranchs, gastropods, ammonites, belemnites, trilobites, crinoids, echinoids, graptolites, vertebrates, plants—even microfossils if you can obtain a slide of some and a microscope). You may find it best to carry out this investigation before reading the relevant section of the text.

(*a*) What is the basic shape of the fossil (ie tubular, flat shells, segmented, globular)? Compare your own description with the technical terms used for each group.

(*b*) What are the major features (eg external ornamentation, internal markings if seen)? Draw 1 specimen and name the features using the diagrams in the chapter to check that you have picked them all out. Add a scale.

(*c*) Look at several specimens belonging to the same group. How much variety is evident in the various features you have just been describing?

(*d*) Discuss the significance and possible functions of the features you have just been drawing and labelling.

(*e*) In which geological period did the fossils you have been drawing and examining live? Show the range of the group on a simple chart.

Investigation 3. Classifying fossils

Take a group of ten different brachiopods or ammonites and construct a key to their classification in the following way:

(a) Select one feature (eg smooth shells or ornamented shells) and divide the group into two parts on this basis. The ratio does not matter: you may have one smooth and nine ornamented.

(b) Select another feature which divides the smooth shells into a further two groups, and repeat this process until each of the ten individuals has been distinguished from the rest.

(c) Complete the same procedure for the ornamented shells.

(d) Record the information on a chart:

	Feature 2A		Individual 1
Feature 1A	Feature 2B	Feature 4A	Individual 2
		Feature 4B	Individual 3
Feature 1B	Feature 3A	Feature 5A	Individual 4
		Feature 5B	Individual 5
	Feature 3B	Feature 6A	Individual 6
		Feature 6B	Individual 7

(e) Compare your classification with one used in a textbook, and modify it to include any new fossils you discover.

NB This process can also be used as an introductory investigation to distinguish between representatives of each of the major fossil groups listed in Investigation 2.

Investigation 4. Fossil evolution

Place a group of ammonites in the order of their appearance in geological time, and show how various features of their shells changed during that time. This can also be done with other groups (eg *Micraster* echinoids).

Investigation 5. Palaeoecology

Examine a large group of fossils obtained from a narrow band of rock.

(a) Do the fossils show signs of abrasion (ie have they travelled far before entombment in the rock, or did they live where the rock was formed)? This will enable you to remove forms which could not have lived in the environment where sedimentation took place.

(b) Divide the fossils into groups possessing similar features and modes of life: ie which were the bottom-livers and which were the swimmers? Is there a wide range of forms?

(c) Take each group and measure the range of sizes included. These can be plotted on a graph.

(d) Draw your own conclusions concerning the difference between the assemblage of fossils and the possible living situation. Use the evidence from the fossils and the sediment in which they are entombed to work out the sort of environment in which these fossils once lived.

Collecting fossils

Fossil-collecting is a fascinating hobby. See how many different varieties you can collect from any exposures of rock near your home (always remembering to obtain the permission of the owner of the land). You may be fortunate and find a rock containing many fossils: the longer you are able to spend on one thin layer the more you will find out about it.

This hobby needs few special tools. Fossils in soft rocks will be prised out easily with a penknife; in harder rocks you will need a hammer and chisel. Many fossils are fragile and must be wrapped carefully in newspaper or sometimes in cotton wool and carried in tin boxes. A notebook is essential to record the exact locality where each one was found, and a small hand lens often helps in the examination of the tinier details.

When you find a layer of rock containing a number of fossils, concentrate on a detailed study of it. Collect samples of each type of fossil, noting their importance in the group as a whole and the place where you found them. It often takes a little time and careful examination before you find the first few fossils in a rock exposure, but the longer you persevere the greater will be your chance of success: you will soon get to know the best places.

You will need a special place to keep your collection tidy for easy reference. When you get home clean each specimen as far as possible so that the details show up, and make neat drawings of all of them. Add your notes on the place where you found each fossil, plotting them all on a diagram or map. If more than one bed is exposed at a place, indicate from which each fossil came. Then look up the name of the fossil in a reference book, which will also give you some assistance in determining its age.

Topics for discussion and essays

(*a*) With the aid of fully labelled diagrams, describe a typical member of three of the following fossil groups:

  trilobites, graptolites, crinoids, echinoids (W).

(*b*) Outline the ways in which fossils can occur and the modes of preservation which are possible (S).

(*c*) Compare the usefulness of the graptolites and the lamellibranchs in stratigraphical correlation (O).

(*d*) Write an essay on 'The importance of fossils to the geologist' (AEB).

(*e*) Considering corals, echinoids, belemnites and graptolites, arrange these groups of fossils in the order of their relative value in stratigraphy, explaining the reasons for your order (L).

(*f*) Explain the points of difference between (*a*) a brachiopod and a lamellibranch and (*b*) a gastropod and an ammonite. Illustrate your answer by careful drawings, giving the name and geological system of each fossil drawn (L).

(*g*) How may fossils be used to determine (*a*) the conditions under which a sedimentary rock was deposited, and (*b*) the geological system to which it belongs? Give the names of some fossil groups that are valuable for these purposes (C).

Fossils are found by looking at the surfaces of fallen blocks of sedimentary rocks. This photograph was taken near Lyme Regis in Dorset.

10 Volcanoes and Igneous Rocks

Volcanic Action Today

Nearly 700 volcanoes are known to have been active since man began keeping records. Some of them are shown on Figure 10.1. Of course this represents a tiny fraction of geological time, and it also ignores what has been happening in the hidden deep oceans, where eruptions still go unnoticed. Those of us who live in Britain have the feeling that volcanic activity is a rare occurrence of little importance, but it has been powerful enough to build the Hawaiian Islands from the floor of the Pacific Ocean. Mauna Loa is higher than Mount Everest with its base 5000 m below the ocean surface, and its top rising to over 4300 m above sea-level. Great plateaus of lava have been formed, and vast quantities of gases (mainly water vapour, carbon and sulphur dioxides) poured into the atmosphere, providing the basic materials for maintaining the oceans on the Earth's surface. We normally hear about volcanoes only in connection with the damage they cause as they erupt, but from a geological point of view we must study them carefully to find out about another aspect of the Earth's past. So many of the rocks we find in the Earth's crust are lavas, or have originated as a molten, mobile mass of rock material, that vulcanology has become a vital part of geology.

We have only one full record of a volcano starting to erupt. Mount Paracutin in Mexico began its life on 20th February, 1943, after two weeks of minor tremors, when a crack opened across a ploughed field and vapours and hot stones were ejected. By the next day there was a pile of debris 10 m high, and this grew to a cone-shaped feature 18 m

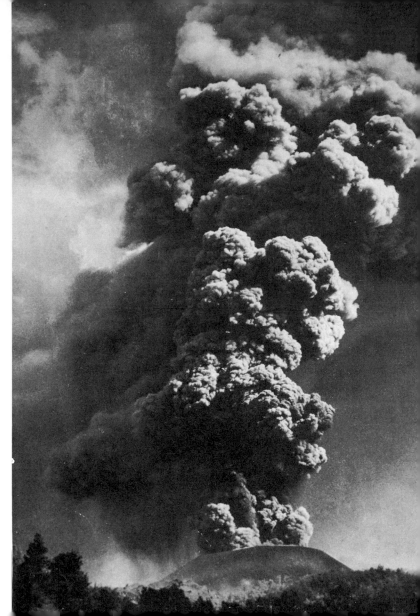

Paracutin in eruption, showing the cloud of dust and gas. (*Ewing Galloway, NY*)

148

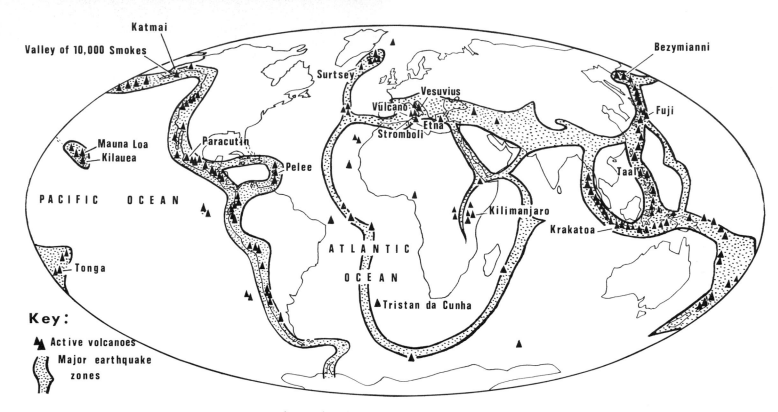

Figure 10.1 Volcanoes and earthquake zones.

high within a week. Lava soon flowed out, inundating local villages, and clouds of ash rose to over 600 m, descending to cover fields and kill trees. In 1946 the cone, composed of lava and ash, was 500 m high, but after that date activity became less intense, and the volcano has been quiet since 1952. The catastrophic speed with which volcanic action takes place is a great contrast to the relatively drawn-out accumulation of sedimentary rocks.

Volcanic activity varies considerably in intensity, depending largely on the type of material being ejected, and on the age of the volcano. Like Paracutin many volcanoes have a period of great activity, but

149

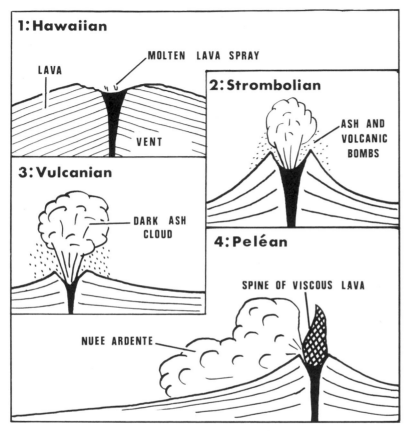

1: Hawaiian

LAVA

MOLTEN LAVA SPRAY

VENT

3: Vulcanian

DARK ASH CLOUD

2: Strombolian

ASH AND VOLCANIC BOMBS

4: Peléan

SPINE OF VISCOUS LAVA

NUEE ARDENTE

Figure 10.2 Volcanic eruptions. Four types of eruption. Note how the intensity of explosion increases as the lava becomes more viscous. In 2 there are minor eruptions every few minutes; in 3 they are more violent, but only occur every few years; 4 is the most violent type of explosion.

then die down and become **dormant** (or quiescent) for many years. The citizens of San José, the capital of the Central American republic of Costa Rica, are having a very difficult time at present because of the eruptions from Mount Irazu 25 km away. Clouds of dust and ash have ruined crops, choked rivers so that they overflow and flood and have caused the city authorities to employ 400 extra road sweepers. The present series of eruptions began in March, 1963, after the volcano had been dormant for many years and life had gone on around it without any interruption. Volcanic peaks where there is no record of any eruption are regarded as **extinct**, but it has been known for such peaks to resume violent activity.

As geologists have studied volcanoes throughout the world they have recognised six main degrees of activity, named after some of the most famous volcanic peaks. Some of them are shown in Figure 10.2.

1 The Hawaiian Islands. These islands are entirely volcanic in origin, formed of very fluid lavas which have been continuously welling out of cracks in the floor of the Pacific Ocean. Many of the volcanoes are in a state of constant activity with lakes of molten lava in their craters occasionally broken by masses of gas causing fountains of incandescent liquid rock as they escape into the atmosphere. Sometimes these lakes rise in level, but at others they sink and cause the surrounding roof rocks to collapse. This forms a wider crater, known as a **caldera**. When the lava rises and pours out of the main volcanic craters, it is partly crystallised and the gases escape leaving a broken, tumbled mass of cindery rock: this is known as **'aa'** lava in contrast to the smoother, rope-like **'pahoehoe'** surface of the more mobile lava issuing from vents lower down on the sides of the cones. Both names are Hawaiian in origin. There are hardly any explosions associated with these Hawaiian volcanoes, because the very fluid lava allows the gases to escape easily.

2 Stromboli. This volcano on one of the Lipari Islands just north of Sicily has a crater which is always full of seething molten lava. This is not so liquid as that issuing from the Hawaiian volcanoes, and there are continuous minor explosions as the gases trapped in the molten material escape. Every ten minutes a mass of lava is thrown into the air, solidifying as 'bombs' which may be up to 15 cm across. This regularity is unusual.

3 Vulcano. Very close to Stromboli is Vulcano, which gave its name to all peaks formed by eruptions from within the Earth. The stiffer (more viscous) lava forms only small flows before they congeal and solidify. A cap of hard rock is soon formed in the crater and stops gases escaping. Eventually the pressure of gas builds up to such an extent that a great explosion takes place, hurling clouds of dark ash into the air. One such violent eruption began in 1888 and lasted for two years. The lava disintegrated in the force of the explosion and was thrown out in large clots which solidified and fell to Earth. At present the mountain is quiet, but further explosions are to be expected from this unpredictable peak.

4 Vesuvius. Although only one-third as high as Etna, Vesuvius is the most famous Italian volcano. We know that there was an exceptionally violent eruption in AD 79, because surrounding towns like Pompeii and Herculaneum were buried instantaneously under ashes or mud, and the top of the volcano was removed altogether by an unusual uprush of gas. Since that date only minor eruptions have taken place, and a new cone is being built up within the old remnant. Activity seems to have increased since 1600, but most of the outbursts have been of Stromboli or Vulcano type.

5 Mont Pelée. The most famous example of the extreme type of volcanic explosion is Mont Pelée on the island of Martinique in the West Indies. In 1902 a volcano, which had not erupted in living memory, exploded and cleared out the top of its vent. Over the next sixteen months a steep-sided dome of lava grew to 300 m high, composed of a very viscous rock, although the solidified 'skin' was often broken by internal explosions. Then quite suddenly one side of the peak opened and a mass of black smoke gushed out. This fiery cloud ('nuée ardente') was composed of droplets of molten lava, ashes and huge boulders of rock lubricated and carried along by expanding gases. It rushed 8 km downhill at speeds of over 300 km per hour to inundate the main island town of St Pierre, where heavy statues were carried several yards by its force and glass was melted. Twenty-eight thousand people were killed as the gases overwhelmed them at temperatures of over 700°C, and then the whole town was buried under ash and lava. The only people to witness this and live were aboard departing ships, for the harbour boiled and many of the boats at anchor were destroyed; one man survived in the depths of the town gaol.

Similar and even greater explosions have been experienced in what are fortunately less populous parts of the world. The uninhabited island of Krakatoa in the East Indies blew away the upper part of its volcanic peak in 1883 with an explosion that was heard nearly 5000 km away. Incandescent clouds of gas and molten lava billowed up 80 km high and were carried around the world by the winds of the upper atmosphere to give exceptionally brilliant sunsets throughout the world during the next three years. More recently, in 1955–6, the volcanic Mount Bezimiannyi in the Kamchatka Peninsula of eastern USSR exploded with great violence and ejected so much rock material that it has been calculated Paris would have been completely covered to a depth of 16 m. A little farther to the east in the Aleutian Islands Mount Katmai was decapitated by an explosion in 1912 when the mingled gases and lava formed a great frothy mass of pumice, which filled a valley 20 km long and 5 km wide, now known as the Valley of Ten Thousand Smokes.

6 Solfatara. The final, declining phase of volcanic action is known as the **solfataric**, after Solfatara, a steam-and-gas vent near Naples. No solid matter is ever thrown out, but mineral-encrusted cones are built up around the gas-vents, or **fumaroles**.

Before we consider the causes of these different types of volcanic activity we must gather some more information from the rocks which are produced.

The Main Structural Features Produced by Volcanic Action

Rocks produced by volcanic action are included in the **igneous** group, because of their association with fiery explosions (*ignis* is Latin for *fire*). Volcanoes erupt solid ashes and volcanic bombs as well as molten lava. Each of these results in distinctive deposits and layers of rock.

Ashes and volcanic bombs are associated with the explosive type of activity, and usually give rise to an **ash cone** round the vent or pipe supplying the material from inside the Earth's crust. Barcena on the west coast of Mexico built a 300 m cone in twelve days in 1952 after a Vulcanian explosion, and the **ring-craters** (maare) of the German Eifel Mountains have each been formed by a single explosion. The fragments involved will include fine ash which has solidified from molten droplets, volcanic bombs from clotted lava, and pieces of country rock torn away from the sides of the vent as the explosion took place (Figure 10.3). The largest fragments will fall back to Earth first,

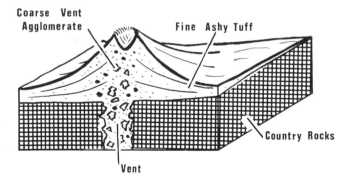

Figure 10.3 An ash cone. Pyroclastic rocks are formed in this way. Note how the deposits become finer farther from the vent, and how easily fragments of the country rocks can be included in the ejected material as they are broken from the walls of the vent.

nearest to the vent. The deposits formed are known as **pyroclastic**. You have probably heard the *pyro-* part used in referring to fireworks which are known as pyrotechnics, or in mention of Indian funeral pyres; *clastic* is a word used for sediments composed of transported rock debris. The main varieties of pyroclastic rocks are **tuffs** in which the grains are sand, silt or clay grades, and **agglomerates** in which the bulk of the fragments are larger. Tuffs are divided into groups according to their composition, and the proportions of crystalline or glassy lava and country rock fragments. **Welded tuffs** and **pumice** result from gas-rich nuée ardente explosions, and resemble lava flows in many respects since they form layers up to 10–13 m thick.

Lava flows are composed of molten rock material issuing from cracks and fissures in the rocks of the Earth's crust, or in smaller quantities from volcanoes. Many are formed underwater in the depths of the oceans, where the gases cannot escape because of the water pressure, and where some of the salts are absorbed from the water resulting in sodium-rich rocks. The tumbled masses of lava which have solidified under such circumstances are known as **pillow-lavas** (Figure 10.4). Lava flows erupted on the land may build up great thicknesses as successive layers accumulate. They vary considerably in thickness, and there may be enough time between eruptions for a soil to form on the surface of the last flow. In the island of Mull, off the west coast of Scotland, the lava flows of Tertiary age are over 2000 m thick. Vast plateaus have been formed by repeated outpourings of liquid lava near Bombay in India (covering 500 000 km²), on the Brazilian Plateau, and in the Columbia and Snake River valleys in north-western USA. These have completely buried the underlying landscape.

When we examine a lava flow in detail we find that the top layer has a rough, slaggy nature with many cavities, or **vesicles**, where gases were trapped during the cooling processes. In older lavas we shall see that these have been filled in by percolating waters and the minerals they carry in solution, like calcite. Many dark-coloured lava rocks are spotted with these small, white, almond-shaped **amygdales**. The lower parts of lava flows, as shown in Figure 10.4, often solidify in **columns**.

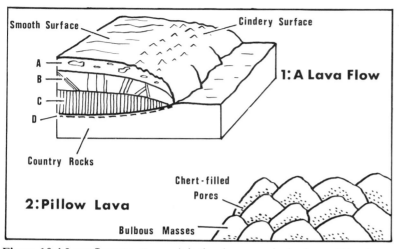

Figure 10.4 Lava flow structures. A is the slaggy, vesicular layer; B is a layer of irregular columns; C is the polygonal column zone; D is the baked section of the underlying rocks.

This is because on cooling the lava contracts, and cracks are formed perpendicular to the cooling surface which isolate columnar blocks. The Giant's Causeway in Northern Ireland is a lava flow which has been partly decapitated by the sea to expose these columns. Another effect of the hot molten lava will be to bake the underlying rocks as it passes over them. The movement of the lava flow may cause the forming crystals to be dragged out so that their longest dimension lies in the direction of movement, and this often leads to distinctive **flow-banding** structures. One of the most important characteristics of rocks

Two lava flows in the cliffs above the Giant's Causeway, Northern Ireland. The columns are at the base of each flow. Notice the difference in degree of weathering between the upper and lower set of columns. (*Crown Copyright*)

153

which solidify from lava flows results from the fact that they cool rapidly in contact with the atmosphere and as they lose their gas content at the Earth's surface. The crystals which form are usually so small that a high-powered microscope has to be used for examining them, and sometimes the cooling has been so rapid that crystals do not have time to form, and a glass is produced. The lavas which were poured out in the recent eruption on Tristan da Cunha were already 70 per cent solid with larger crystals that had formed during the passage of the lava, and these were soon made fast by the rapid cooling of the remaining molten material in a fine-grained mass.

When the lava is ejected through a small, circular vent, rather than a long fissure, it will spread out in a circular feature and cool as a **cone-shaped volcanic hill**. If the lava is fluid and at a high temperature it will be able to flow over considerable distances before solidifying, and will form **shield volcanoes** with low-angled slopes, like those in the Hawaiian Islands. Stiff, viscous, cool lava will solidify almost as soon as it reaches the surface, and will form steep-sided **domes** and **plugs** growing by additions from inside, since the solid outer layers cannot be penetrated. These are common in the Auvergne district of the French Central Massif.

Most of the larger volcanoes are **composite**, being made up of both ash and lava. The initial explosion and eruption of ash has been followed by the outpouring of lava which impregnates the ash and forms a more resistant landform. Whilst ash cones are soon destroyed by river action, the composite cones stand out in the landscape forming many of the best-known volcanic peaks like Fujiyama (Japan), Vesuvius (Italy) and Kilimanjaro (Tanzania).

Finally, **geysers** and **hot springs** are also the result of volcanic action. They are caused by the effect of hot volcanic substances on groundwater, superheating it until the expanding steam gushes out through any tiny crack. Any minerals caught up with the rush of steam are deposited round the geyser mouth. If calcium carbonate is dominant the deposit is known as **travertine**; if silica as **sinter**.

The Main Structural Features Formed by Igneous Rocks Trapped beneath the Surface

Although we only see what is going on at the surface, a large proportion of molten igneous material is trapped in the rocks of the Earth's crust long before it reaches the surface and solidifies there. Such rocks are known as **intrusive**, in contrast to the **extrusive** types which form the volcanoes and lava flows (Figure 10.5).

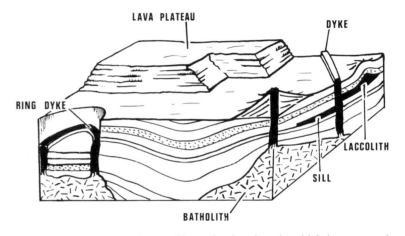

Figure 10.5 Igneous landforms. The main situations in which igneous rocks are found, and some of the surface features they produce.

One important group of intrusions occurs in large sheets, only a few metres thick, but extending over many kilometres and often found together in great numbers, or swarms. Many of these sheets push their way along bedding planes in between layers of rock, but others take any cracks in the rocks, like faults, and it is thought that sometimes the molten rock forces its way through the Earth's crust by permeating and replacing the rocks in its way. **Dykes** are the sheets which cut

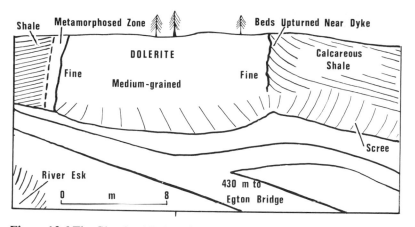

Shale | Metamorphosed Zone | Beds Upturned Near Dyke

DOLERITE

Fine | Medium-grained | Fine | Calcareous Shale

Scree

River Esk

0   m   8

430 m to Egton Bridge

Figure 10.6 The Cleveland Dyke. The outcrop of the dyke near Egton Bridge, Eskdale, North Yorkshire. This exposure illustrates the variation of grain-size within a dyke, and the effect on the surrounding rocks. Write up your own description of this feature.

across the other layers, as the diagram of the Cleveland Dyke in Figure 10.6 shows, whilst **sills** are those which push in along the bedding planes and are more or less parallel to the other layers of rock in the area. A **laccolith** is a thickened, lens-shaped sill mass, often formed of more viscous rock. Dyke swarms are common in the islands along the west coast of Scotland, and are sometimes found in cone-shaped sheets along conical fractures, and sometimes in parallel swarms. The Isle of Arran is crossed by over 600 dykes which have widened the island by 2 km. Sill sheets are also common in this island, but one of the most continuous is the Whin Sill of northern England, which causes a series of waterfalls in the upper Tees Valley, is the foundation for part of Hadrian's Wall, and eventually reaches the Northumberland coast at Bamburgh Castle and the Farne Islands. Most sheet intrusions have solidified more slowly and have larger crystals than the rocks of lava flows, except on the margins, where

chilling took place most rapidly in contact with the country rocks (Figure 10.6). Whereas the lava flows have a lower chilled margin and upper slaggy layer, the minor sheet intrusions have chilled margins on both sides, an important point for distinguishing between lava flows and sills in a group of rocks.

The more massive intrusions tend to cool even more slowly because they do so at greater depths, and because the retention of the gases trapped in a large mass of rock material helps to lower the rate of heat loss. They are characterised by very large crystalline grains, easily seen and distinguished. Because these intrusions solidify at depth they are known as **plutons**. The largest **batholiths** may be several hundreds of kilometres long and up to 150 km wide, and usually push the surrounding rocks aside to such an extent that they are folded, and altered considerably by the heat of the igneous mass. Large fragments of country rock may be broken off in the course of the intrusion, and may form definite patches in the batholith (**xenoliths**), or may be completely 'digested' by melting. Dartmoor, and most of the granite masses in Britain, are batholiths which have been exposed at the surface after millions of years of erosion of the overlying rocks. As the weight of the overlying rocks is gradually released in this way joint cracks open up in the granite in addition to those formed as the rock cooled; the later group are usually parallel to the margins of the intrusion. It has recently been shown that the series of granite masses in the South-west Peninsula are connected at depth, and it is probable that Dartmoor, Bodmin Moor, Land's End and the Scilly Isles are smaller offshoots (**stocks**) of a large, still uncovered batholith.

In some places molten rock has been forced upwards from its deep-seated reservoir by the collapse of a section of the crust, especially in the vicinity of volcanic action. The process is known as **cauldron subsidence**, and happened on the Isles of Mull and Arran. A great slice of crustal rocks subsided, bounded by almost circular cracks, up which the molten material was forced to solidify as **ring-dykes**.

The Origin of Volcanic Activity and Igneous Rocks: the Magma
The impressive and often catastrophic results of volcanic activity have

occupied many scientists in a consideration of their causes. Unfortunately the means for studying the origins of such activity are not within our grasp, as are those for studying the origins of the sedimentary rocks, for all igneous phenomena are the result of deep-seated processes which we cannot observe.

When we look at a map of the world (Figure 10.1) we can see that volcanic igneous activity is at present confined to certain areas closely associated with earthquakes, and that large areas of the world are free from both. Even within these major zones there are differences amongst the lavas erupted from the volcanoes. It seems therefore that there is no world-wide reservoir of igneous material, but that each area is subject to local melting of the rocks near the base of the Earth's crust, and that there is a certain amount of modification of the molten material as it migrates upwards through the crust. If this is so, we do not really understand how the melting takes place, since temperatures at the base of the crust are probably 500° or 600°C, which is well below the melting point of igneous rocks even at the surface. We have to assume that some abnormal extra heating takes place locally, and causes the solid rock to become a molten and gaseous **magma**. This magma migrates towards the surface along cracks in the rocks, and is sometimes trapped on the way.

Magmas are silicate melts which vary a great deal in composition. **Acid** ones contain a great deal of silica (65–75 per cent) and are rich in alkalis (potash and soda) but deficient in iron, lime and magnesia. **Basic** ones have much less silica (45–55 per cent), little of the alkalis but a good deal of iron, lime and magnesia. **Alkaline** magmas, besides having a lot of soda and potash, have about 60–65 per cent of silica. Much more basic magma than acid or alkaline reaches the surface. It does so at higher temperatures (1000–1200°C) and is much more fluid, whereas acid varieties are cooler (800–900°C), and more viscous and explosive: this factor accounts for many of the features of volcanic activity. The gases contained in a magma are extremely important, and may be equal in volume to the liquid constituents. The gases supply the motive force. Cotopaxi is a volcanic peak on top of the Andean ranges in South America, but the pressure of gases ascending 7000 m above sea-level was still enough to hurl a 200-ton block of rock 16 km in 1929. The most intensive volcanic explosions like those of Mont Pelée, Katmai, Krakatoa and Bezimiannyi are accompanied by such great volumes of gas that the magma becomes a froth and eventually disintegrates into a spray as the gases expand in their upward rush.

As the magma cools, be it in a surface lava flow, or in an intruded mass, it solidifies. The minerals contained in the molten mass crystallise at different temperatures. Those minerals that are rich in magnesium, iron, calcium and titanium tend to have high melting points and so tend to crystallise before minerals rich in silicon, potassium, sodium and aluminium. In a basalt, for instance, the order is often:

(1) Magnetite—largely iron, no silicon.
(2) Olivine—mainly magnesium, iron, little silicon.
(3) Augite—calcium, magnesium, iron, plus medium amounts of silicon.
(4) Plagioclase—calcium, sodium, aluminium, plus more silicon.

The first minerals to crystallise float freely in the remaining melt. They may be heavy and sink, forming a zone near the base of the igneous body rich in that mineral, or lighter and float to the top. This process is known as **differentiation**. The remaining molten magma may react chemically with the first crystals, and if there is sufficient time these crystals will be completely replaced. **Zoned crystals** are formed when the process is not completed and involve mineral groups where some change is possible in their composition without affecting the basic structure: the best examples are the plagioclase feldspars, in which each zone has slightly different proportions of sodium, calcium, aluminium, silicon and oxygen. In other cases reaction with the residual magma may lead to the formation of overgrowths of other minerals, as in the example of **Corona structure** shown in Figure 10.7. The order in which minerals replace each other in this way is known as the **reaction series** (Figure 10.7). The two main branches involve the ferro-

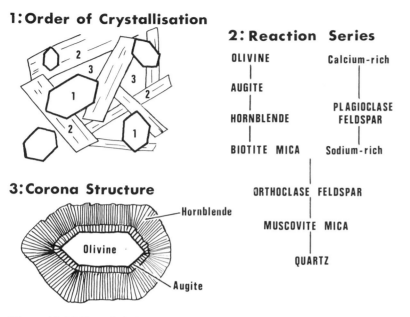

1: Order of Crystallisation

2: Reaction Series

OLIVINE   Calcium-rich
|
AUGITE
      PLAGIOCLASE
      FELDSPAR
HORNBLENDE
| |
BIOTITE MICA Sodium-rich

ORTHOCLASE FELDSPAR

MUSCOVITE MICA

QUARTZ

3: Corona Structure

Hornblende

Olivine

Augite

Figure 10.7 Minerals in igneous rocks. Some of the crystal relationships are shown in 1 and 3, and the general order of cooling reactions in 2. The minerals in the Corona Structure increase in silica content towards the outside.

magnesian and feldspar minerals. As cooling proceeds further new crystals form round the first group, giving rise to a mesh of almost solid rock, and finally the last minerals fill in the interstices that are left (Figure 10.7). We can usually tell the order of crystallisation by examining the shapes of the minerals which make up an igneous rock. The first crystals to form have good crystal faces (**euhedral**); the second group are formed round the first (**subhedral**); and the last fill in the gaps and have no recognisable shape (**anhedral**). When all the main minerals have crystallised there may still be fluids and gases circulat-

ing through the rock, and the last stages in the cooling of a magma involve the reactions of these fluids and gases with the minerals that have already formed, and their eventual crystallisation in cracks to form distinctive **mineral veins**. These latter stages are most common in the deeply buried plutonic rocks, because the gases escape easily from lavas.

The cooling processes may be interrupted. A magma may have been cooling slowly, forming large crystals, as it moved upwards towards the surface, but may finish the series of events rapidly as it comes into contact with the atmosphere. This will result in a matrix of very fine crystals between the larger ones formed in the early stages of cooling: the example of the lavas on Tristan da Cunha has already been quoted. A rock of this type is known as a **porphyry**, and the larger crystals it contains as **phenocrysts**. On the other hand two magmas may meet and mix, or some country rock may be absorbed and alter the composition of the magma. This process of **contamination** is important, like that of differentiation, in giving rise to the many varieties of igneous rocks.

The General Characteristics of Igneous Rocks

Before we begin describing the different types of igneous rocks we are likely to find in the field, it will be a help to summarise the main grounds of comparison and contrast. Each rock will have a distinctive chemical composition, mineral composition and textures, and all of these aspects have a special significance, since they are closely linked with the way in which the rock was formed.

Chemical Composition. The chemical composition of igneous rocks is dominated by **silica**, which makes up between 40 and 75 per cent of the total in its important position as the basis of the silicate group of minerals. Igneous rocks have been divided according to the quantity of silica they contain, but this is not obvious when examining hard specimens and so we do not recommend such a procedure.

If they contain over 65 per cent silica they are **acid**;
if they contain 55–65 per cent silica they are **intermediate**;

157

if they contain 45–55 per cent silica they are **basic**;
if they contain under 45 per cent silica they are **ultrabasic**.

Other important chemicals are the oxides of aluminium, iron, magnesium, calcium, potassium and sodium. Water is common in magmas, but usually escapes from lavas as steam, though it may be trapped in deep intrusions and be absorbed in micas and other special minerals there.

Mineral Composition. This is largely dependent on the type of magma and on its history of cooling. Each rock has a few minerals, upon which we base its name, and a host of others in very small quantities. The important group are the **essential** minerals, and those of less significance are the **accessory** group.

Rocks rich in silica (acid) are also rich in potassium, sodium and aluminium, and the minerals forming will be rich in these elements, ie quartz, alkali feldspar (orthoclase and sodium plagioclase) and the micas. Hence the rock will be light in colour and weight.

As the proportions of the oxides of silicon, potassium, sodium and aluminium decrease those of calcium, magnesium and iron oxides increase, and in the basic rocks (low silica) the minerals to form will be rich in calcium, magnesium and iron, and poorer in silica. These are the heavier and darker minerals (the ferro-magnesian group—olivine, augite, hornblende), plus calcium plagioclase, and so the rocks will be heavier and darker.

Textures. The textures of igneous rocks also reflect the ways and places in which they were formed. The grain-size, for instance, is indicative of whether the rock cooled on the surface (glassy or fine-grained), as a minor sheet intrusion (fine or medium-grained), or as a major plutonic intrusion (medium or coarse-grained). Other textures, such as porphyritic crystallisation, flow structures and the shapes of the individual crystals, give us further knowledge concerning the cooling history of the rock.

These are the main factors we use to 'pigeonhole' the different types

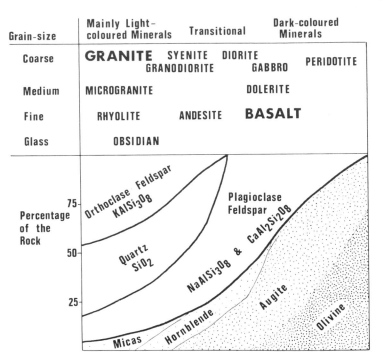

Figure 10.8 Igneous rock classification. A classification of the igneous rocks, based on mineral composition and grain-size. The darker-coloured minerals are shown in the bottom right-hand corner of the chart.

of igneous rock. Over 600 varieties have been named, but these can be placed in the larger groups shown on the chart in Figure 10.8. Four igneous rocks are shown in detail in Figure 10.9. As we examine the main groups of igneous rocks you will find that a very important group, the granites, are not mentioned in this chapter. We cannot discuss them until we have noted the characteristics of metamorphic rocks, and so they are left until the end of the next chapter.

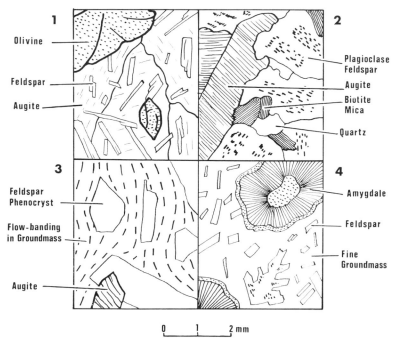

Figure 10.9 Some igneous rocks. These are four sketches as you might see the rocks under the microscope. Note their interlocking crystalline nature, and the varying sizes of the crystals and the relationships between them. See if you can give a name to the four rock-types illustrated here (answers at the end of the chapter).

The Basic Igneous Rocks: Basalts, Dolerites and Gabbros

Eighty per cent of the lavas poured out on the Earth's surface are basalts. Dolerites are the most common rocks occurring in minor sheet intrusions, but gabbros are relatively rare plutonic rocks.

Basalts. These are black, heavy rocks found in lava flows, and sometimes on the margins of minor intrusions. They are the most common type of lava rock, and it has been estimated that more than 5 million km³ of basalt has been extruded from the Earth's interior in the last 180 million years. The crystals in basalts are too small to be seen in the hand specimen of rock, which has a dull, broken surface. The only minerals which stand out are the phenocrysts, and amygdales may also be conspicuous. Basalts are essentially formed of the calcium-rich plagioclase feldspar (labradorite) and augite, but they are usually divided into two major groups depending on the proportion of another ferromagnesian mineral, olivine. The first consists of rocks which contain **essential olivine** (up to 20 per cent); these **olivine basalts** are particularly common in the Pacific Ocean islands such as Hawaii. Some varieties in these areas contain even more olivine because of differentiation which took place in the magma during cooling. The plateau-forming basalts of continental regions are known as **tholeiitic basalts**, have very little or no olivine, and sometimes even contain a small amount of quartz. A third group of basalts is formed beneath the sea as pillow-lavas, and the rocks are known as **spilites**. The plagioclases have reacted with the seawater, and have been altered to varieties richer in sodium.

Dolerites. Dolerites are similar in composition to the basalts, but have larger crystals, which can just be distinguished in hand specimens and give the rock a small-scale mottled appearance. These rocks occur in sills and dykes. Like basalts they are dominantly dark in colour, and also have a related variation in composition—some are **olivine-dolerites** and others are **quartz-dolerites**. Porphyritic dolerites are common with phenocrysts of olivine, augite or feldspar. The dykes and sills formed of dolerite are very common in areas dominated by basic igneous rocks, and their total volume adds up to thousands of cubic kilometres. Most of our British examples of minor intrusions (eg Whin Sill, Cleveland Dyke) are formed of dolerite, and the southern half of the Isle of Arran is riddled with dykes and sills of many varieties of dolerite. One of the most famous sills in the world is the Palisade Sill of New Jersey, USA, which forms steep cliffs looking across the Hudson River to New York. It is an interesting example of the process of differentiation, and the result is illustrated in Figure 10.10.

and each occurrence gives rise to a unique rock-type. Differentiation has led to a layered structure in many of these plutonic intrusions. One large gabbro mass on the east coast of Greenland has layers of **olivine-gabbro** at the base, whilst the topmost horizons are formed of **quartz-gabbro**.

The Bushveld Complex of South Africa is also layered, as Figure 10.11 shows in a simple summary, but is on a much larger scale: it is

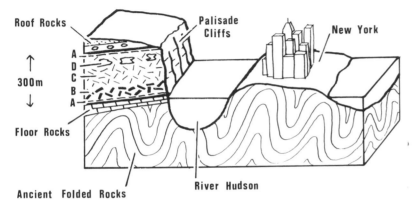

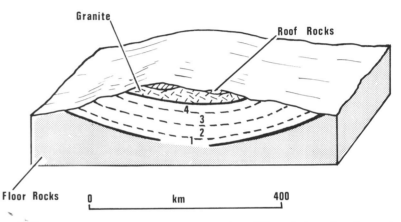

Figure 10.10 The Palisade Sill. This sill forms the cliffs overlooking New York across the Hudson river. Which part of it is a cumulate rock, i.e. has been formed by differentiation and the sinking of heavier olivines through the remaining melt? Zone A represents the fine-grained, chilled margins; Zone B is olivine-dolerite; Zone C is coarse dolerite; Zone D is patches of coarse-grained rock rich in feldspar and acidic minerals.

Figure 10.11 The Bushveld Complex. How has differentiation played a part in the formation of different rocks within the lopolith? Zone 1 is the basal zone of chilled rocks; Zone 2 is the Critical Zone, up to 2 km thick, of gabbro with veins of chrome-rich ores; Zone 3 is the Main Zone, over 5 km thick, of gabbro with veins of titanium-iron ore; Zone 4 is the Upper Zone, 3 km thick, of red rocks such as syenite and diorite.

Gabbros. Gabbros are the coarse-grained equivalents of basalts and dolerites, and like the latter sometimes have a mottled appearance due to the pale feldspars. The large black crystals show up, even in a completely dark variety, because of their shiny cleavage faces. The gabbros show even greater variety than the other two basic groups,

the same size as Scotland, and several kilometres thick. For some time it was regarded as a saucer-shaped sill type of intrusion, known as a **lopolith**, but this is now disputed. The weight of the rock has caused the area to sag, and the complex is regarded as a whole series of intrusions, rather than as one in which differentiation took place to enrich the floor rocks in heavier minerals and in which contamination took place in the upper layers.

British gabbros are mostly to be found in north-east Scotland and in Skye (the Cuillin Hills), where the composition varies because of differentiation and contamination: in places large xenoliths of shales have been caught up into the igneous mass. This latter process results in patchy rocks containing unusual minerals like garnet.

The Ultrabasic Igneous Rocks

There is a graded group of rocks from the gabbros, which contain a high proportion of feldspar, to the ultrabasic rocks with a very low percentage of silica and scarcely any feldspars. They are composed almost entirely of the heavy, dark, ferromagnesian minerals like olivine, augite and hornblende, and are always coarse-grained. Rocks rich in olivine are known as **dunites**, and those dominated by a mixture of olivine and augite as **peridotites**.

Ultrabasic igneous rocks are a small group occurring sometimes at the base of gabbro intrusions, and containing mineral ores such as chrome, platinum and nickel. At other times they are found in larger masses in mountainous areas, and it is thought that the intense movements during mountain-formation led to the incorporation of rock material from beneath 'Moho'. The investigations which have been made of earthquake shock records indicate that the upper Mantle is formed of a substance like peridotite.

Other Lava-forming Rocks: Andesites, Rhyolites and Trachytes

Although basalts form most of the lava rocks, some of the less important types are very distinctive and must have a place in any consideration of igneous activity.

Andesites. The andesites are closely related to basalts. The main differences are due to a higher percentage of silica: the plagioclase feldspar has slightly more sodium in it (andesine), hornblende may be as important as augite, and quartz is more often present in small quantities. Most andesites are almost as dark as basalt and equally fine-grained;

they may be porphyritic and often contain amygdales. The two rocks are difficult to tell apart in hand specimen. Andesite is second in importance to basalt as a lava rock, and is especially characteristic of the mountain-building regions around the margins of the Pacific Ocean: its name comes from the Andes Mountains of South America. It occurs with both basalt and rhyolite lavas in these areas. The mountain-building movements must have caused large-scale contamination of the original basic magma with crustal material, because the intermediate and acid varieties are more common than the basic rocks.

Rhyolites. Rhyolites are the most acid group of lavas, and come to the surface in a viscous state. Rapid cooling leads to the formation of a glass, or even to pumice, and these acid lavas are commonly associated with tuffs. Snowdon, the highest mountain in Wales, is built of a pile of largely acid lavas and tuffs of Ordovician age over two km thick. It is now recognised that many acid lavas and tuffs were formed in great Peléan-type explosions, and they are known as **ignimbrites**. The main minerals in the acid lavas are quartz, orthoclase feldspar, sodium-rich plagioclase feldspar, and tiny quantities of mica and hornblende. The glassy varieties are **obsidian**, which is shiny black with a conchoidal fracture and common flow-banding, and **pitchstone**, which contains more water and has a greenish, resinous appearance. Both of these varieties may contain phenocrysts and the tiny beginnings of crystals, which were 'frozen' as they formed and are known as crystallites. The acid lava rocks occur in small quantities in a variety of conditions, and are most probably formed from the local melting of rocks in the upper layers of the Earth's crust as they are affected by large basic intrusions. The molten mass migrates to the centre of igneous activity to form a distinctive group of rocks. Pitchstone often forms minor intrusions, such as the sills along the coast of Arran just south of Brodick.

Trachytes. These are alkaline lavas, moderately rich in silica and characterised by their richness in the alkalis so that orthoclase and soda-rich plagioclase are abundant. They are often grey in colour and commonly porphyritic. Such rocks are found around the East African rift valleys, where they must have been influenced by the Earth's movements. The discovery of unique carbonatite lavas rich in minerals like calcite, which do not normally occur in igneous rocks, in this area, has led to the conclusion that these rocks have been formed from a special magma ascending from great depths. Elsewhere it is evident that there has been contamination of such rocks by limestone, with a consequent lowering of the silica percentage.

The **medium grained equivalents** of these rocks, which sometimes occur as lavas, and more often in minor sheet intrusions, or on the margins of plutonic intrusions, are of small importance compared with the dolerites. They are often porphyritic, and may be known by the main phenocryst mineral (eg **quartz-porphyry**). In other cases they are related to the coarse-grained member of similar composition (eg **microgranite**). Igneous rocks of similar mineral composition, with only differing grain-size to separate them, are often grouped together in 'clans', named after the coarse-grained variety: thus granites, rhyolites and microgranites are part of the **granite clan** of rocks.

Patterns of Igneous Activity

We come to the point where we can summarise the evidence thrown up by our studies of igneous rocks and the features they produce. There are at least three particular conclusions arising from this evidence: they concern the origin of the igneous rocks, their involvement in the major processes affecting the evolution of the Earth's surface and the part they play in working out more detailed, local Earth history.

(1) The most fundamental conclusion is that igneous activity transfers molten rock, known as magma, from a position at greater depth into the crustal rocks, or even to the surface. Surface activity is manifested in volcanic outbursts.

(2) The scale of volcanic activity varies: there are local cones, eg Vesuvius near Naples in Italy, which involve a relatively small quantity of material; there are the vast volcanic domes, eg the Hawaiian Island group, which include thousands of cubic kilometres of volcanic rock; there are the extensive basalt plateaus, such as the Deccan plateau in the north-west of the Indian peninsula and the Snake river plateau of north-western USA, which are composed of hundreds of thousands of cubic kilometres of igneous material; and there are the ocean-floors, which are now seen as the products of igneous activity and cover two-thirds of the Earth's surface. When all these forms of activity are added together it is seen that even the surface volcanic effects alone contribute approximately $3 \cdot 5$ km^3 per year to the crustal rocks; taken over the millions of years of geological time this adds up to immense quantities. When we consider that we live on a planet which is gradually using up its own internal sources of energy, we must at least allow for the fact that volcanic activity has been of greater importance in the past.

(3) The distribution of igneous activity at the present moment, as shown on Figure 10.1, is confined largely to narrow zones which are also characterised by earthquakes: these are the mobile zones of the Earth's crust, which are separated by much larger areas where disturbances of this nature are few. Compare this map with Figure 6.5 showing the distribution of ocean-floor features, and the folded mountain belts of most recent origin. Most of the volcanic action is connected with particular sites: the oceanic ridges, the island arcs on the continental sides of the ocean trenches and the ranges of young folded mountains. Other sites of volcanic activity include a more widespread distribution over the ocean-floors, rift valleys on the continents and the major basalt plateaus—often near the continental margins.

The characteristic oceanic vulcanism is non-explosive. The ocean ridges seem to be the site of continuous upwelling of tholeiitic basalts: these push apart the ocean-floor on either side as they are intruded between walls of rock which have just cooled and solidified. The many volcanic islands and submerged volcanic peaks (seamounts) are sometimes associated with the ocean ridges (eg Iceland, which straddles the ridge in the North Atlantic, its new neighbour Surtsey, and the Azores farther south), but are also more widely scattered. These scattered ocean volcanoes include the Hawaiian mass, rising over 10 000 m from the ocean-floor. Whilst the lower horizons are normally tholeiitic basalt lavas, upper layers, erupted later, may include trachyte and rhyolite.

Another distinctive volcanic association includes the young folded mountains and island arcs. Volcanic rocks are found amongst those involved in folded mountains: basalt pillow lavas erupted on the seafloor where sedimentation was taking place, piles of andesite lavas like those associated today with the island arcs, and ignimbrites resulting from highly explosive activity taking place into the atmosphere. Since uplift these regions have been the continuing scene of a variety of volcanic activity, involving rock-types which range from fluid basalts to highly explosive, silica-rich varieties. In addition these regions contain larger bodies of igneous material, trapped before they could reach the surface. Granites are particularly important in volume, and their relationship with the deeper seated mountain-building activities is discussed more fully in the next chapter. There are also masses of gabbro, and smaller intrusions, ranging from dolerite to microgranite and silica-rich porphyries.

Other continental regions subject to volcanic activity are the rift valleys, like those in East Africa, where an unusual variety of rocks is produced. Large volcanoes, like Mount Kilimanjaro, and piles of alkaline basalt lava (Ethiopia) are interspersed with smaller volcanoes built of rare carbonatite rock, rich in carbonate minerals. The large basalt plateaus, erupted from elongated fissures in the crustal rocks, form a contrast with the rift valley areas, and show closer relationships with the oceanic lavas.

(4) These distributional features, and the rock-types involved, suggest that the igneous rocks are formed as follows. Almost all magmas must originate near the base of the Earth's crust, where the rocks are of

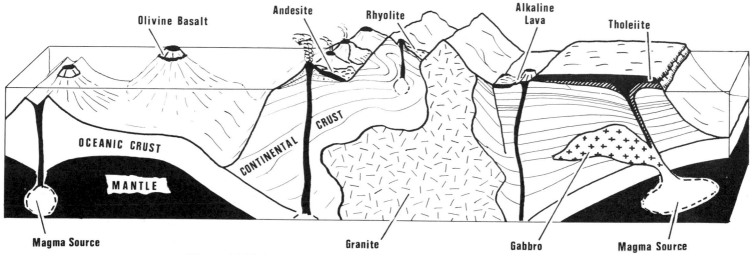

Figure 10.12 A pattern of igneous activity. Relate the types of igneous rock to the situations in which they occur. Contrast the igneous rocks produced by activity in the ocean areas, along the continental margins and in the continental interiors.

basaltic type, rich in olivine. Rocks which have solidified from this magma without any modification are normally only found in oceanic areas, where they have not had to flow through great thicknesses of crustal rocks. The tholeiites have probably suffered a small amount of contamination during their progress through the crust, and andesites, involved in intense mountain-building movements, are contaminated to a greater extent by the crustal rocks, which tend to increase the silica-content of the magma. The reasons for alkaline lavas being associated with the rift valleys is not clear, but there is evidently some relationship with an unusual, deep-seated magma. Acidic lavas are produced from local melting and the formation of a migrating magma from crustal rocks rich in silica, like sandstones. On rare occasions melting takes place at exceptional depths, incorporating ultrabasic rock material into magmatic fluids. All these phases of igneous activity are illustrated in Figure 10.12.

Differentiation is most common in the larger intrusive bodies of basic magma, and leads to several distinctive layers within a single intrusion. Most of the large intrusions, however, are of acidic granite, and their origin is considered in the next chapter.

Igneous Rocks and Time
Igneous rocks become incorporated with a whole series of sediments and metamorphic rocks once they have solidified, and an important task for the geologist is to examine their relationships in time with the surrounding rocks.

The **extrusive lavas and tuffs** can be treated like sedimentary rocks, since they were formed on top of an older layer, and were covered by a younger bed. Lava flows are usually chilled at the base (resulting in smaller crystals), and have a slaggy or weathered upper surface (Figure 10.13.1).

Intrusive rocks were formed after all the rocks which they affect by contact (Figure 10.13.2). They have chilled margins on both sides, and may also have baked the surrounding rocks. We can thus discover the earliest date for the period of intrusion, but the latest date at which it could have taken place may be less easy to place. In the situation

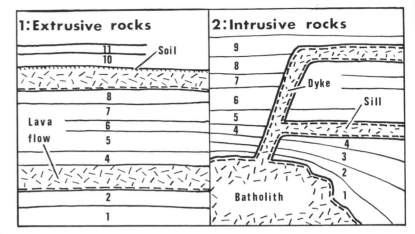

Figure 10.13 Igneous rocks and time. Rocks 1–11 are numbered according to age: 1 is the oldest and 11 the youngest. The igneous rocks are dated by comparing them with these numbered sediments. Note that the lava flows are only chilled at the base in A; but both margins of the intrusions in B are chilled. The sedimentary rocks would be heated and altered where they meet the chilled margins.

depicted in Figure 10.13.2 the upper sill is younger than rock layer 9, but it may be equal in age to layer 10, or to any layer formed at a later date. It is most helpful if a dyke is cut across by erosion and then covered by later rocks: then we can say that the dyke is older than the rocks which cover it.

Igneous Activity and Mineral Wealth
Few of the common rock-forming minerals are of value to man, but igneous activity also brings quantities of other minerals which are in demand for their usefulness or beauty: metallic **ores**—rocks contain-

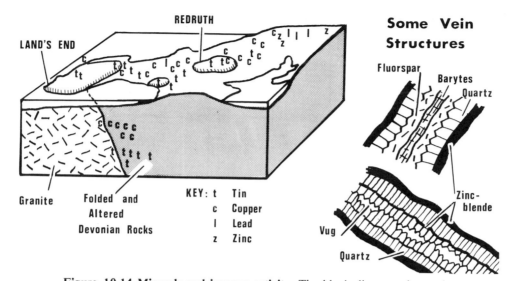

Some Vein Structures

Figure 10.14 Minerals and igneous activity. The block diagram shows the arrangement of ore-bearing veins around the Cornish granites. Can you detect any zoning—which ores are nearest the granites? The upper vein structure is crustification; the lower comb-structure. Which are the gangue minerals?

ing minerals which are exploited on a commercial scale for the extraction of the metals they contain—and **gemstones** are two forms of mineral wealth commonly associated with igneous origins.

Whilst the metallic elements, such as iron, tin, copper, lead and zinc, form a small proportion of the total crustal volume, they are often concentrated locally by geological processes. The ordinary rock-forming processes of cooling magma seldom produce such concentrations, but there are cases of local enrichment by differentiation: iron minerals settle to the bottom of a cooling melt and may accumulate to such an extent that it is worth mining such rocks. Gabbroic intrusions are often rich in iron for this reason: the famous Swedish iron ores of Kiruna may have been formed in this way, but in general such deposits are not common.

Most metallic ores occur in veins which have formed after the main mass of igneous rock has solidified: hot gases and fluids permeate the rocks of the surrounding country, and the minerals they contain are deposited after chemical reaction with these rocks (**metasomatism**), or as the **hydrothermal** fluids cool so that certain minerals are precipitated out of solution. Figure 10.14 shows the relationships of this type of mineral deposit to the Cornish granites. The veins so formed are composed of the metallic ores (eg cassiterite containing tin, galena with lead, or zincblende) and other minerals known as the **gangue** (eg fluorspar, barite, calcite, quartz: see Chapter 3 for descriptions). Some of these gangue minerals, which used to be discarded by the miners, have recently become of value, and old mine tips are being re-worked.

Whilst dealing with metallic ores we must also mention that many are formed by non-igneous processes. Some are formed after weathering has enriched an original deposit (Figure 4.5), or when deposition occurs in limestone cavities (page 32) or river or coastal placers (page 92). Ironstones are often of sedimentary origin, formed by offshore accumulation. Metamorphism (see next chapter) may also lead to the enrichment of the metallic content of a sediment, thus making it worth mining. A combination of processes will also lead to a sufficient concentration, and the origin of some of the world's largest ore bodies is extremely complex. The iron ores of the Lake Superior region in North America were originally banded Precambrian sediments of a type which are common all around the world, but the iron content of 15–40 per cent would not be enough to warrant extraction: surface weathering or metamorphism can raise this percentage to 55–60 per cent iron (the Labrador deposits, those of Cerro Bolivar in Venezuela, Minas Gerais in Brazil, and Krivoi Rog in USSR—amongst the world's largest—are similar in origin).

Many problems are associated with the mining of ores and other economic minerals. Opencast pits and quarries take away valuable farming land; underground removal of rock can lead to subsidence and damage to housing, or the formation of areas of flooded land at the surface. More scientific planning has sought to reduce these dangers in recent years. Figure 8.12 shows how one such scheme is applied.

Gemstones are crystals which are rare and resistant both to corrosion and abrasion. They also have high refractive indices and can be cut so that they sparkle. Many of the colours in them are due to rare impurities occurring in normally colourless crystals.

Diamonds are the most valuable gemstones, and for many years South Africa has been the main producer, although Russia is now the scene of the majority of new discoveries. The diamonds often occur in river gravels, into which they have been washed from unusual volcanic

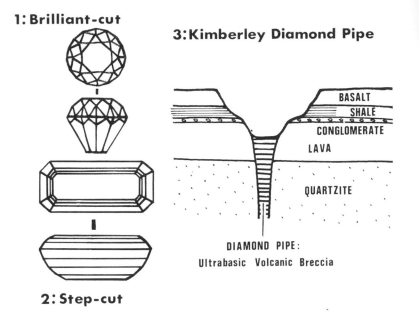

Figure 10.15 Diamonds. Diamonds are cut as shown in 1; rubies, sapphires and emeralds as in 2.

pipes filled with blue-ground, or kimberlite, an ultrabasic igneous rock. The most famous Kimberley pipe has been mined to nearly 1300 m and is illustrated in Figure 10.15. The rock has to be crushed to extract the diamonds, and a huge volume of rock has to be examined for each stone of value. The beauty of a diamond is largely due to its high refractive index, which is used to advantage in the 'brilliant-cut' treatment (Figure 10.15). Since diamond is also the hardest substance occurring naturally, the poorer quality stones are used for cutting hard metals, and for studding the drill-bits used in rock-borings.

The other main gemstones, and their characteristics, are listed in Figure 10.16.

MINERAL	COMPOSITION	HARDNESS	DENSITY	COLOUR	MODE OF OCCURRENCE	MAJOR PRODUCING AREAS
I. PRECIOUS STONES						
DIAMOND	Carbon, C	10	3·52	Colourless	Volcanic pipes; river gravels	South Africa; Congo; USSR
RUBY SAPPHIRE }	Corundum, Al_2O_3	9	4	{ Red (Chromium impurity) { Blue (Titanium)	Metamorphic rocks; gravels	Burma; Siam; Ceylon; Kashmir
EMERALD AQUAMARINE }	Beryl, $Be_3Al_2Si_6O_{18}$	7.5	2·7	{ Green { Blue-green	Metamorphosed limestones Pegmatite veins near granite	Colombia; Ecuador; Peru South America; Asia
II. SEMI-PRECIOUS STONES						
TOPAZ	$(AlF)_2SiO_4$	8	3·5	Usually yellow, brown	Pegmatite veins with tinstone	Brazil; USSR
TOURMALINE	Boro-silicate of aluminium	7·5	3·2	Yellow/green/blue/pink/red	Pegmatite veins	Brazil; USSR; USA
GARNET	Complex silicate	7	4·2	Green/pink/red/yellow	Metamorphic rocks	Widespread
PERIDOT	Olivine, $(Mg,Fe)SiO_4$	7	4·2	Green	Dark-coloured igneous rocks	St John's Island (Red Sea)
ZIRCON	$ZrSiO_4$	7·5	4·7	Colourless (cf. Diamond)/green/red/yellow/blue	Light-coloured igneous rocks, river gravels (gems)	Ceylon
SPINEL	$MgAl_2O_4$	8	3·7	Red/sometimes blue	With corundum in metamorphic rocks	Burma, Ceylon
QUARTZ	SiO_2	7	2·6	Colourless: ROCK CRYSTAL Purple (Manganese): AMETHYST Brown (Iron): CAIRNGORM Pink (Titanium): ROSE QUARTZ	Widely distributed in the rocks; gems from pegmatite veins	Brazil; Uruguay; USSR; India; Ceylon
TURQUOISE	Phosphate of aluminium, copper	6		Green-blue	Amorphous cavity fillings	Persia (in lavas); Mexico; USA
LAPIS-LAZULI	Silicate of sodium, aluminium	6·5	3	Ultramarine	Metamorphosed limestone	Afghanistan
JADE {	Nephrite; silicate of magnesium, calcite Jadeite: silicate of sodium, aluminium	6·5	3·3	Green Green	Nephrite needles in schists Jadeite in altered feldspars	China; Turkestan Burma

Figure 10.16

THINGS TO DO AND DISCUSS

Investigation 1. Igneous rocks

Examine specimens of igneous rock (basalt, dolerite, gabbro, andesite, obsidian, pitchstone—and granite when you have read the next chapter) and ask yourself similar questions to those we suggested in connection with the sedimentary rocks. If possible have a look at some of these rocks under the microscope in thin section and try to recognise the minerals contained. Make neat notes of your descriptions in your practical books.

Investigation 2. The chemical composition of igneous rocks

Plot the following figures on a graph as shown, and answer the questions:

Oxide	Peridotite	Gabbro	Diorite	Syenite	Granodiorite	Granite
SiO_2	43·54	48·36	51·86	59·41	66·88	72·08
Al_2O_3	3·99	16·84	16·40	17·12	15·66	13·86
Iron oxides	12·35	10·47	9·70	5·02	3·92	2·53
MgO	34·02	8·06	6·12	2·02	1·57	0·52
CaO	3·46	11·07	8·40	4·06	3·56	1·33
Na_2O	0·56	2·26	3·36	3·92	3·84	3·08
K_2O	0·25	0·56	1·33	6·53	3·07	5·46
Others	1·83	2·38	2·83	1·92	1·50	1·14

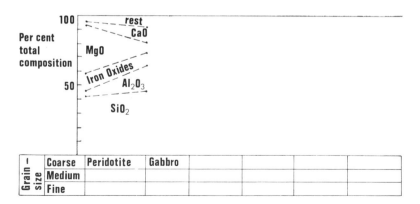

(1) Describe the variation in silica (SiO_2), iron oxides, and the oxides of calcium, sodium and potassium.

(2) Compare the chemical compositions of the rocks with the mineral compositions as shown in Figure 10.8.

(3) The rocks mentioned here are all coarse-grained varieties. Add the names of the medium- and fine-grained varieties to the graph.

Investigation 3. Find out about recent volcanic activity

Make notes on the type of feature produced, the nature of the rocks poured out and the distribution of such activity. Frank W Lane's book, *The Elements Rage*, gives several descriptions, and the Smithsonian Institute, Washington DC, USA, sends details of all current Earth phenomena to interested schools.

Investigation 4. Model volcanoes and lava flows are described in Vols 1 and 2 of *Geology*, the journal of the Association of Teachers of Geology.

Topics for discussion and essays

(a) Explain briefly the difference between volcanic rocks and other igneous rocks. Describe the principal characteristics of three of the following: gabbro, porphyry, rhyolite, obsidian (W).

(b) Write a short essay on the forms of igneous intrusions which can now be seen due to erosion (S).

(c) Give an account of the main types of acid igneous rocks with examples of localities where they are found (S).

(d) Comment on the statement that the nature and the form of volcanoes are related to the composition of their products (O).

(e) Write a short essay on the crystallisation of magma (O).

(f) Make a simple classification of basic igneous rocks. Then select two of the rock types and describe them fully (composition, textures and structures) (L).

(g) Make a list of the common minerals that make up the bulk of the igneous rocks. Describe the properties of any four of them and state the rocks in which you would expect to find the minerals you describe (C).

(*h*) Distinguish between ore minerals and gangue minerals. Give concise mineralogical descriptions of three ore minerals and two gangue minerals (L).

(*i*) Write an essay on either sedimentary iron ores, or hydrothermal mineral deposits (AEB).

(NB The four rocks in Figure 10.9 are:

1 Olivine dolerite

2 Gabbro

3 Pitchstone with flow-banding and phenocrysts

4 Amygdaloidal andesite)

11 Metamorphic Rocks and Granites

What Is Metamorphism?

Metamorphism means 'change'. Many rocks have been subjected to different conditions of pressure and temperature from those in which they were first formed, and this has led to the formation of new minerals, new rock textures, and even completely new rocks with no traces of the original. Any previously existing rock may be affected—sedimentary, igneous and even rocks which have been metamorphosed already.

Both metamorphic and igneous rocks are **crystalline** in nature, but the fact that distinguishes them is that whilst the igneous rocks are formed from molten, flowing magmatic material, the metamorphic varieties are recrystallised in place, and whilst still largely solid: they do not migrate at all. Some rocks are mixed in the sense that they are partly altered by reaction with migratory fluids. Granites are found to have relationships with surrounding rocks which suggest that they can be either metamorphic or igneous in origin. In Finland the granites commonly pass gradually into intensely metamorphosed rocks, and were obviously the result of the same phase of alteration. The Dartmoor massif in Devon, however, is equally clearly an intrusive body of granite, which has pushed aside the surrounding rocks but only produced a narrow zone of alteration caused by the original heat of this igneous mass. There is a striking case in the Bergallo massif in the Alps, where both aspects are combined (Figure 11.1). The deep-seated fusion of crustal rocks to form the granite was followed by upward injection of the relatively light rocks. We shall return to this subject at the end of the chapter: it is mentioned here to illustrate the essential differences between igneous and metamorphic rocks.

Metamorphism takes place because of changes in temperature,

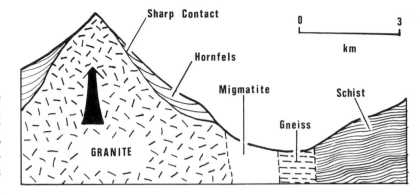

Figure 11.1 Granites. The granite at the base of the mountain has typical metamorphic relationships, but that at the top is obviously intruded and is therefore an igneous rock. This is part of the Bergallo massif in the Alps.

because of increasing pressure and because of chemical reactions caused by migrating fluids. **Temperature changes** may be the most important factor locally, and thermal metamorphism is particularly effective when rocks are in contact with large masses of molten igneous rock, or when they are buried deeply. Rocks take millions of years for any visible changes to take place below 200°C, but as the temperature rises above this figure the reactions become more and more vigorous until parts of the rock become fluid at some stage over 700°C. The minerals contained in the final metamorphic rock give a guide to the temperature at which the alteration took place. The other main agent of metamorphism is **increasing pressure**. As the confining, or hydrostatic pressure increases due to sideways compression, or deep burial, reactions within the rock are encouraged and minerals having their atoms 'packed' more densely are favoured. Garnet is the most common example of such new minerals formed under extreme pressure. Rocks affected by directed pressure as in dynamic metamorphism have their minerals reorientated with their long axes at right angles to the directions of pressure. Recent laboratory experiments have shown that certain rocks under pressure adapt themselves to more confined spaces, whilst others shatter. Finally the **fluids** percolating through and permeating the rocks near igneous intrusions and deeper in the Earth's crust may affect the rocks and accomplish considerable chemical alteration.

Some Features of Metamorphic Rocks

Some metamorphic rocks are **granular**, having almost equal-sized grains interlocking, but without any sign of being arranged in bands or in parallel fashion. Banding, or **lineation**, is a common feature of many metamorphic rocks, in which **cleavage**, or **foliation**, has developed as the minerals are reorientated with their long axes parallel to each other. This is best developed in **schistose** rocks, with their narrow bands of flaky micas. The difference between the two groups is caused by the relative absence or importance of directed pressure in the act of metamorphism.

The actual grain sizes in metamorphic rocks depend at first on the sizes of grains in the original rocks; thus there are fine-grained slates as well as coarse-grained marbles. As the processes become more intense the coarser textures become dominant and very large crystals may form, known as **porphyroblasts**. They may even retain some of the structural features of the surrounding minerals from which they were formed, such as the fine laminations continued in quartz inclusions across a garnet porphyroblast.

The type of metamorphic rock that results from the various changes is very closely related to the geological circumstances in which it is produced, such as nearness to a large intrusion of hot, molten, igneous rock, or involvement in a zone of mountain-building, or deep burial in a sagging geosyncline. We shall consider the different rock types in these connections.

Contact or Thermal Metamorphism

The heat generated by an igneous intrusion affects a wide band of country rock around it, and metamorphic changes take place within this zone, or **aureole**. It may vary in width from a few millimetres near minor intrusions to several kilometres, and the amount of alteration decreases as the distance from the intrusion increases. The contact with the hot molten body effects many of the changes, but the fluids and gases invading the rocks around the intrusion also cause widespread alterations. Pressure plays a very minor role.

The most common type of rock produced under these circumstances of intense heat and little pressure is the **hornfels**, a coarsely granular (ie little or no mineral alignment) type of rock in which all trace of the original structure is lost. It grades outwards into partially altered, 'spotty' rocks containing clots of new crystals. The fine-grained sedimentary rocks are altered most by this process, and clearly demonstrate the progression from fine- to coarse grain size, the increase in hardness, and the higher temperature minerals as the intrusion is approached. As Figure 11.2 shows, the gabbro intrusion at

The Insch Gabbro

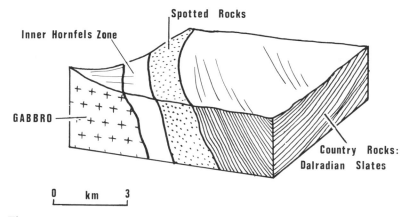

Figure 11.2 Contact metamorphism. A generalised diagram showing the metamorphic aureole around the Insch gabbro intrusion in north-east Scotland.

Insch in north-east Scotland has affected rocks which had already been subject to mild metamorphism in which many of the original clay minerals had been altered to chlorite and white mica. The spotted zone of the new aureole contains tiny masses of dark biotite mica, chiastolite and cordierite crystals, whilst the innermost zone, next to the gabbro, is a coarse hornfels composed of quartz, feldspar, cordierite and andalusite crystals. If the metamorphism had been at higher temperatures another mineral, sillimanite, would have been formed; if lower temperatures had prevailed, the 'cordierite zone' might have been next to the igneous mass. A series of such mineral zones has been recognised, but one has to examine the rocks under the microscope in the laboratory to determine which minerals are present.

Orthoquartzite sandstones and acid igneous rocks are probably affected least by contact metamorphism. The sandstones, largely composed of quartz with a little feldspar, may be recrystallised to form a compact and resistant **metaquartzite** which will have lost most of the original structural details like current-bedding or ripple-marks, though these are sometimes preserved by lines of mica flakes. Less pure sandstones have new minerals after metamorphism. Acid igneous rocks are not affected much, because the renewed high temperatures resemble closely those at which they were formed. Basic igneous rocks, however, alter more readily because the temperatures at which metamorphism takes place are usually lower than those at which they formed and new minerals, stable at those temperatures, come into being.

Limestones recrystallise to white, sugary looking **marbles** if they are almost pure calcium carbonate. Tiny amounts of impurities may cause red or green streaks, but these white rocks are the only true marbles in spite of the fact that many people give that name to any polished rock. Impure limestones often contain some silica, alumina, iron and magnesia and when metamorphosed develop a whole group of new minerals such as diopside (one of the pyroxenes) and tremolite (a fibrous amphibole).

Although some sedimentary structures may be preserved such as cross-bedding in the quartzites, and graded bedding in schistose grits, it is rare for fossils to be found in the metamorphic rocks.

Dislocation or Dynamic Metamorphism

The enormous stresses set up by the directed pressures in Earth movements cause rocks to be fractured along fault lines, or to be compressed into folds (cf Chapter 12). The breaking of rocks often causes a wide zone to be shattered, and under moderate pressures this forms a breccia (cf Chapter 8). Increasing pressure leads to the rock being broken down into a fine dust of microscopic particles, which may became mobile in extreme cases. These 'crush-rocks' are known as **mylonites**, and involve considerable recrystallisation, as well as the distortion of the more fragile minerals.

Cleavage planes at right angles to the direction of compression are formed when fine-grained rocks are subjected to intense pressure, as illustrated in Figure 11.3. **Slates** are dull grey rocks which may have a greenish or mauvish tint, and they cleave along fine parallel planes.

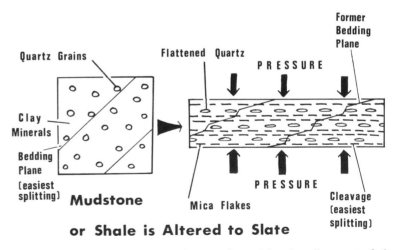

Figure 11.3 Cleaved slate. The cleavage formed by the alignment of the minerals enables the rock to be split along planes parallel to this direction which is perpendicular to the direction of maximum pressure exerted on the rock.

This made them in great demand for roofing in the past, because they are also tough. Most of the roofing slates came from great quarries in the Snowdonia area of north Wales, where the sediments and lavas, which had otherwise been unmetamorphosed by the mountain-building movements at the end of the Silurian period, were squeezed up against the resistant block of Anglesey. The fine Cambrian shales were altered to slates in this process.

Regional or Dynamothermal Metamorphism
Whilst contact and dislocation metamorphism are caused by intense heat or pressure affecting the rocks of a relatively small local area, regional metamorphism involves the deep burial of large masses of rock, and its results can be seen over widespread areas. Regional metamorphism is usually associated with mountain-building, where the rocks are buried as a portion of the Earth's crust sags, and are subjected to both high temperatures and intense pressures. The products vary according to the depth of burial, and various intensities of action can be distinguished.

1 **Low-Grade Regional Metamorphism: Phyllites.** This involves considerable pressures, but at relatively low temperatures. The resistant rocks, like sandstones, are gradually altered to more compact rocks with an interlocking granular structure. Many metaquartzites or marbles formed in this way show evidence of the stresses in deformed crystals and a degree of schistosity. As in contact metamorphism the fine-grained shales and mudstones, containing minerals stable at low temperatures (eg the clay minerals), are most affected, and are usually completely recrystallised to phyllites containing new flaky minerals (mostly chlorite and mica). They are brittle and have a surface sheen. Organic matter is broken down to specks of graphite or cubes of iron pyrites.

2 **Medium-Grade Regional Metamorphism: Schists and Granulites.** Rising temperatures cause the final relics of sedimentary and igneous structures to be lost as the rocks become more plastic and develop strong foliated structures. The fine-grained sediments form mica-schists with very thin alternating layers rich and poor in mica. As the degree of alteration increases porphyroblasts become more common, and garnet-mica-schists, having large red crystals in the shiny micaceous rock, are often found. Basic igneous rocks are also recrystallised, and hornblende-schists (greenschists) or amphibolites result; but even at this stage the resistant sandstones take on only a very rough schistose nature. The quartz grains of these rocks recrystallise in an interlocking mesh and granulite rocks are formed. British examples of these rocks can be seen in Anglesey and in the Highlands of Scotland.

3 **High-Grade Regional Metamorphism: Gneisses.** The most intense metamorphism causes the rocks to become plastic with free recrystallisation. In the earlier stages the recrystallisation took place within the framework of an almost solid rock, but under the most extreme conditions the rock is completely transformed. Gneisses are coarse-grained, roughly banded rocks and often have alternating streaks of dark- and light-coloured matter. The darker layers have a greater degree of schistosity, being composed of hornblende and mica, and the lighter part is more massive quartz and feldspar. Gneisses often grade into granites, and are associated with migmatites (described below).

When the regional metamorphism takes place large areas of many square kilometres are affected, but there are usually patches of more intense metamorphism in the centre and a gradation outwards to low-grade rocks. Thus there may be a change from fine slates on the margins to schists containing garnet porphyroblasts, and then to coarse gneisses in the centre. As in contact metamorphism there is a series of recognisable zones.

Regional metamorphism is associated with mountain-building, but some areas show evidence of more intensive alteration than others. The Snowdonia area of north Wales has already been referred to as an area of low-grade action, and a similar result was produced by folding at a later date in Devon and Cornwall. Schists and granulites predominate in the Scottish Highlands. It is only the very ancient 'shield' areas of Scandinavia, northern Canada and Africa that are largely

A **quartz vein** cuts across the contorted, banded gneiss. (*Crown Copyright*)

made up of gneiss and granite. Two metamorphic rocks can be examined in detail in Figure 11.4.

Metamorphic Rocks and Time

The date of a period of metamorphism is normally assessed by working out the dates of the rocks affected and the overlying sediments. The separating unconformity provides a major line of reference in such studies. In north-west Scotland the most ancient rocks are highly metamorphosed schists and gneisses, and are covered, above an unconformity, by Precambrian sediments. The metamorphic rocks are thus of early Precambrian age. The alteration of the aureole around the Dartmoor granite must have taken place after the Upper

174

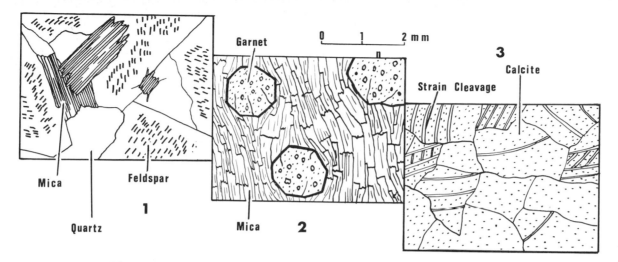

Figure 11.4 Granite, schist and marble. These are three microscope sections, showing the characteristics of the minerals contained in the rocks: 1—granite; 2—garnet-mica-schist; 3—marble. Can you explain the differences between the shapes of the mica grains in 1 and 2? What is the difference between 3 and a sedimentary limestone (Figure 8.10)?

Carboniferous period (the age of the rocks affected by metamorphism), but as there are no rocks of later date in contact with the granite, or lying near it without any alteration, it is difficult to give a final date for the ending of the metamorphism.

Polymetamorphism means that an area has been subjected to more than one phase of alteration, and that these can be detected from a study of the minerals and other structures contained in the rocks. Geologists who make studies of ancient, altered rocks are dependent on the preservation of distinctive minerals, and there are often cases where an area is affected by a second phase of metamorphism similar in grade to the first and this results in few, if any, changes. Some parts of the extreme north-west of Scotland show evidence of several phases

of regional metamorphism and it is supposed that there have been as many periods of mountain-building. One example of this is shown in Figure 11.5, which is a good illustration of how the structures of complex rocks can be sorted out, and can give us a lot of information about the history of an area.

If a region of gneiss rocks is affected by low-grade alteration the rock will be modified, but usually some of the high-temperature minerals will remain, and the process is known as **retrograde metamorphism**.

Migmatites

Migmatites are highly metamorphosed rocks which have an added

(1)

(2)

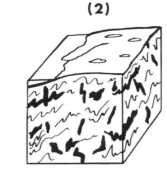

Dykes

Folded,
Altered Rocks

Figure 11.5 Polymetamorphism. 2 is a block diagram illustrating the rock structures in some of the oldest British rocks, the Lewisian schists and gneisses of north-west Scotland. The most recent period of folding and metamorphism broke up the layers of rock, reconstructed in 1: these had already been folded and altered at least once.

component. This has resulted from the migration of magmatic fluids circulating through the rock pores in an extremely fine state. Most of these rocks have a streaky or patchy appearance, and are characterised by fold structures which testify to highly plastic conditions at the time of formation. The actual rock-type depends to a large extent on the original 'host-rock'. Thus former sandstones and shales will be richer in quartz and mica than the basic igneous rocks, which will have more darker patches. All these rocks, however, with the addition of the acid quartz–feldspar mixture, become richer in potassium feldspars.

The migmatites occur mostly in zones just outside the most intensely metamorphosed regions, where the effect of the migrating fluids is obvious, but has not dominated and completely altered the rock by being involved in the highly plastic deformations such as produce gneisses or granites.

Granites

Granites and the related granodiorites are twenty times more important than all the other plutonic rocks. They have always been regarded as igneous rocks, and are included in the classification of the igneous rocks adopted in this book (Figure 10.8). The fact that granites are nearly always associated with areas of mountain formation, and are often an integral part of a group of regionally metamorphosed rocks, has caused lengthy disputes to arise concerning their true position and origin.

All granites are similar so far as rock characteristics are concerned. They are all largely formed of quartz and potash feldspar, with some plagioclase, biotite and muscovite. When the plagioclase is a little more abundant and a little more calcic, and the amount of biotite increases, the rocks should strictly be called granodiorites—which are actually commoner than true granites. Granites are mostly coarse-grained and granular, the grains being of roughly the same size and having irregular outlines: it is often difficult to distinguish which cooled and crystallised first. Occasionally a complex interlocking of some crystals shows that they became solid at the same moment. In hand specimen granites are amongst the easiest rocks to identify, as the crystals are large and stand out plainly. The dominating white or pink feldspar (up to 75 per cent of the volume of some varieties), the greyish-looking but glassy quartz (usually about 20 per cent of the rock volume), and the shiny, black mica grains, are all distinctive. The appearance of a granite under the microscope can be seen in Figure 11.4. Some granites are porphyritic, like the variety from Shap on the eastern edge of the Lake District. This rock has large pink phenocrysts of orthoclase feldspar set in a mass of small crystals of white plagioclase feldspar, grey quartz and black mica, and is quite unique: it is of particular use when found as a glacial erratic.

When most of the granitic material has crystallised there are still mineral-rich liquids and gases circulating through the deeply buried mass, and these finally solidify in a distinctive series of deposits which form veins across the granite or in the surrounding rocks. **Aplites** are medium- or fine-grained granitic veins, and **pegmatites** are very coarse-grained. Exceptionally large crystals of mica and feldspar have been obtained from pegmatites, which must have continued to be mobile for a long time. Crystals of orthoclase the size of a house have been found in Norway, and micas up to 3 and 5 m across are reasonably common in South Africa. The reasons why aplites and pegmatites differ are not fully understood, but they seldom occur together. Other liquids permeate the surrounding rocks, and often form mineralised veins containing copper, tin, tungsten and other metal ores (Figure 10.14). The final stages of granite-formation may be accompanied by the alteration of the feldspar minerals by gases. The kaolin (china clay) of the granite moors near St Austell in east Cornwall was formed in this way.

The main distinctions between granites are not so much in their composition and rock characteristics, as in the circumstances in which they are found. Some are closely associated with metamorphic rocks, since they grade into migmatites and banded gneisses; others are just as certainly intruded into a group of other rocks which they have pushed aside and altered by heat action; and in many cases the granites have made room for themselves by replacing the original rock (stoping). Nearly all the granites, however, are associated with areas of mountain-building in the recent or more distant past.

It seems that the metamorphic granites were the original granites, formed in such conditions of migmatic injection, heat and pressure, that a series of crustal rocks was completely broken down to a state where they became virtually molten and mingled together to form a rock composed of quartz, alkaline feldspar and mica, having a granular crystalline texture, but no banding, as a result of flowage, pressure or partial injection. We cannot tell what the previous rocks were like. It is possible for any type of rock to be altered to a granite after millions of years of additions of granitic material until the host is eventually unrecognisable. As granitisation takes place the incoming fluids react with the rock, leaving the acid, granitic material and removing any basic constituents. These processes take place in rocks which have been dragged down to great depths in basins of sedimentation known as geosynclines (cf Chapter 12). This phase is followed by the uplift of the rocks into great ranges of folded mountains, and as this happens there are large-scale migrations of the granitic material, which is relatively light, into the rising mountains where they are emplaced as forceful or permitted (eg ring-dyke) intrusions. This is how most intrusive granites originated, but some granites have been found outside areas of mountain-building. Such granites are part of a series of volcanic rocks resulting from basic magmas, and are probably formed at a late stage of activity as the residual magma (after differentiation has removed all the basic constituents) solidifies.

Diorites and **syenites** are two other rocks found in plutonic associations, but only occur in minor quantities compared with the granites, or even the gabbros. Diorites are akin to the andesites in composition; syenites to the trachytes. They are most probably formed by the contamination of granitic or basic magmas, and many of these rocks can actually be traced into large masses of granite from which they are offshoots. They are coarse-grained, and can be distinguished from granite only by the absence of quartz in hand specimen. Syenites often have pinkish feldspars, whereas diorites usually have white plagioclases and a larger proportion of dark minerals (eg hornblende or augite).

The Rock Cycle: a Summary of Rock-forming Processes
In your studies of the common rock-types occurring at the Earth's surface you may have been struck by the fact that there are only three major groups; or you may be feeling overwhelmed by the many varieties within these groups; or you may still be trying to distinguish between, say, an igneous and a metamorphic rock. Whatever your position as you reach the end of this section of the course, you must realise that all the processes of rock formation are part of normal Earth activity and result from the reactions of rock material to different environments. Thus rocks formed at high temperatures and pressures inside the Earth often become unstable and crumble easily in the face of low temperature and pressure conditions when exposed to contact with the atmosphere.

Perhaps it is most important to realise that all rocks are rather temporary links in an extremely long chain of geological evolution. A

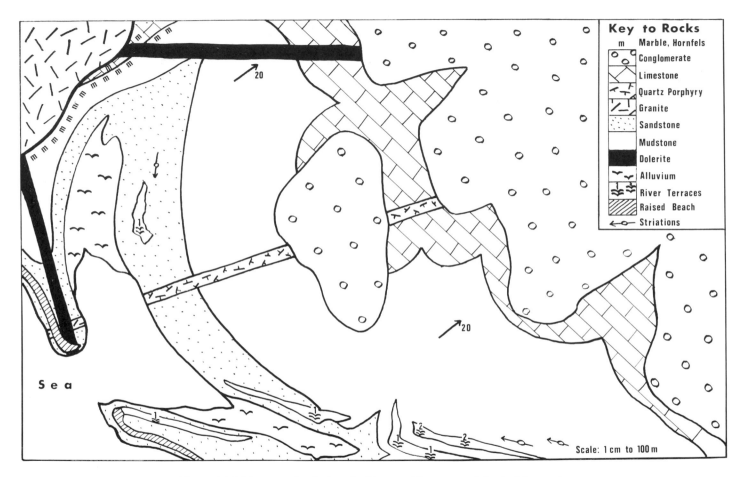

Figure 11.6 Map exercise. Relate the history of igneous activity in this area to the sequence of sedimentary rock formation. Comment on the distribution of metamorphic rocks in the area. How can you tell that this area has been subject to the effects of a glaciation, and of a fluctuating sea-level?

lava flow erupted at the surface will be weathered and eroded; the debris will be transported to a site of deposition where it becomes a sedimentary rock; it may then be buried deeply and metamorphosed into a slate or schist; and then it may be uplifted as part of a mountain chain and re-exposed to weathering. This type of recycling of rock materials goes on and on, in the ways shown in Figure 11.7. New material is brought up from below the Earth's crust (ie from the zone known as the Mantle) to give rise to new basalts and gabbros, and adds to the material forming the continental masses. Some rock material is also lost for long periods to the ocean-floors. But most continues to be recycled, and this may involve physical and chemical breakdown and

reconstitution. Minerals like quartz may last through several of these cycles: rounded, frosted, wind-deposited sand grains may be re-deposited in the sea with little change of character. A mineral such as olivine, however, soon breaks down to clay minerals; calcium plagio-clase may dissolve into constituent ions and the calcium ions may recombine with carbonate ions to form the shells of marine creatures, which become incorporated in a limestone. The elements composing the minerals combine and recombine as the rock cycle progresses, just as do the minerals which make up the rocks.

Building and Roadstones

A wide variety of rock-types have been used as **building stones**. For many ages the nearest compact, load-bearing stone which could be quarried in squarish blocks was used for normal buildings. Today little natural stone is used anywhere in Britain, since cheap bricks and concrete are easily transported anywhere. Even where natural stone is used it is largely as a decorative, facing material, rather than as the load-bearing part of the building.

A study of older buildings suggests that almost every locality had its source of building stone. The best are often known as freestones— ie they could be cut or split in any direction: some of the Permian red sandstones and the Jurassic oolitic limestones are of this nature. Gran-ites have been used extensively where hard-wearing properties are valued: kerbstones in the older sectors of our towns and the old London Bridge were built of Dartmoor granite. At times the best building stones were not available, and a second best had to suffice— or else the inhabitants of the area used local clays, not of any use as a building stone, for brick-making. The villages along the northern mar-gins of the Vale of Pickering in Yorkshire were close to Jurassic oolites and the older houses were built of this material, but the villages along the southern margin, separated by marshes from the oolites, used the local chalk for their buildings.

Roofing materials have also changed with the years: the slate quar-ries of north Wales have been put out of business by the manufacture of clay and concrete tiles. Before the slate boom in the nineteenth

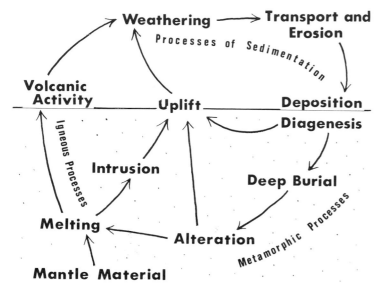

Figure 11.7 The rock cycle. A summary of the rock-forming processes. Describe the history of a quartz grain from the time of its formation in a granite, through its incorporation in a greywacke to its present outcrop in a gneiss.

179

century houses were roofed with the nearest thin-bedded rock material: this may have been a sandstone several centimetres thick. Failing the presence of such materials—which also necessitated the construction of strong walls and cross-beams—thatch was the answer.

Much rock is now extracted and crushed: it is then used as road aggregate, or recast into building blocks for interior walls or even for decorative outer finishes.

Roadstones require a number of properties: they should be resistant so that they will not wear away rapidly and will not dissolve; and they should be able to bind easily with the tar. The most important stones used today for road surfaces are the Palaeozoic limestones and hard, silty sandstones, together with medium- and fine-grained basic igneous rocks. Shales are excavated for road bedding. At the Cliff Ridge Wood quarry, near Great Ayton, Yorkshire, a dolerite dyke was excavated many years ago for roadstone, leaving a great trench across the landscape, and now the hardened shales on either side are being removed for bedding.

THINGS TO DO AND DISCUSS

Investigation 1. Metamorphic rocks
Examine specimens of metamorphic rocks, including hornfels, slate, marble, schist and gneiss, and make a special study of granite. Write up the results and description in your practical books.

Investigation 2. Summary: rocks of the Earth's crust
Make summary lists of the rocks you have studied under the headings, (a) Origin, (b) Composition, (c) Uses.

Investigation 3. Building and roadstones
What evidence can you find of their extraction in your own neighbourhood? If buildings are being erected in your town centre, inquire about the materials being used, and where they come from.

Topics for discussion and essay-writing
(a) What is meant by metamorphism when the word is used in petrology? In what geological settings are you likely to find (i) a metamorphic aureole and (ii) foliation? (S).
(b) Write brief notes on the nature and origin of three of the following: quartzite; mylonite; marble; garnet-mica-schist; cordierite-hornfels; hornblende-gneiss (O).
(c) Describe the changes when a mudstone is (i) intruded by a large igneous mass, (ii) affected by regional metamorphism (O).
(d) A piece of country is described as 'composed of metamorphic rocks'. Discuss what this tells you about the geological history of the area and mention the rock-types you might expect to find there (L).
(e) Which minerals are the common constituents of granite and how would you recognise them? How would you distinguish a granite from (i) gneiss, and (ii) sandstone? (N).
(f) Briefly describe the nature, mode of occurrence and economic uses of three of the following: (i) chalk, (ii) slate, (iii) gypsum, (iv) galena, (v) marble (N).

12 Structures in the Rocks

So far our examination of the evidence available in the rock record at the Earth's surface has been confined to those features which were formed at the time when the rock itself was laid down or solidified. We have studied the composition, grain-size and internal structures associated with the sedimentary, igneous and metamorphic rocks, together with the remains of ancient life trapped in the sedimentary rocks.

Most sediments and lava flows began as more-or-less horizontal layers, but when we start to investigate rock outcrops in cliff faces and quarries we discover that it is unusual to find horizontal rocks: they are normally tilted, bent or fractured, and often criss-crossed by many cracks. A study of these structures provides us with information concerning the post-formational history of the rocks: their reaction to the stresses and strains which act on them as part of the Earth's crust, and to uplift. We refer to such structures as of tectonic origin (Gk, *tecton* = constructor, builder), in order to distinguish them from the depositional or cooling structures.

Studying Tectonic Structures
Certain observations can be made at rock outcrops.

1 **Are the rocks the right way up?** Are the oldest at the bottom and the youngest at the top as we would expect, or has the whole group been turned upside down? How can we tell? There are a number of points to consider.

The presence of unconformities is always helpful, because the younger beds cut across the eroded ends of the older (Figure 8.21). Minor structures in sedimentary rocks, such as false-bedding and graded-bedding (Figure 8.2), ripple-marks, rain-pits and sun-cracks give us the most useful evidence concerning the 'way-up' of a group of rocks.

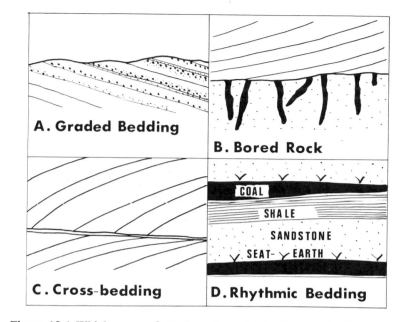

Figure 12.1 Which way up? Each section of this diagram, A–D, contains evidence which will enable you to determine the original sequence in which the rocks were formed.

Figure 12.1 shows how a knowledge of these facts helps a geologist to work out the structure of an area. When lava flows are included in a group of rocks we can look for an upper weathered or vesicular zone and lower columnar layer (Figure 10.4). In coal-bearing rocks we can look for roots going down below the coal seam, but absent from the overlying strata.

2 **What are the strike and dip of the rocks?** Having determined whether the rocks are the right way up, or have been completely overturned, we shall want to define their position. We must be sure we

have an actual bedding surface, and not a cross-bedding lamination, metamorphic cleavage plane, fault or joint surface. We also need a clear surface on which to take measurements: all rock surfaces must be defined in terms of a three-dimensional space, and a flat, vertical quarry face will not give a true reading.

Two directions at right angles to each other are used to define the attitude of a bedding plane. Of these the **strike** is the easiest to determine: it is the bearing of a horizontal line across the surface. The instruments we use to measure the strike include one which will enable

us to measure the horizontal (a level or clinometer), and one which will give us a bearing (a compass): the two may be combined in a single instrument. Once the strike bearing has been established a line can be scratched on the rock surface. The **dip**, which is the angle of maximum slope, or inclination, from the horizontal, can then be measured in a direction at right angles to the strike by the clinometer: it may be expressed as, say, 30 degrees towards the south (Figure 12.2).

A similar approach can be adopted when analysing a geological map (see also Chapter 20). Strike lines can be drawn at different heights across a bedding plane: in the case of Figure 12.2 they would be east–west. They can also be regarded as structural contours—ie lines drawn across a bedding plane (or other structural surface such as a fault plane) which join all places of the same height above or below sea-level. In this form it is most helpful to relate them to the relief contours, as is done in Figure 12.4.

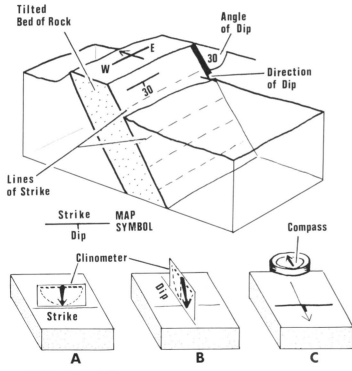

Figure 12.2 Strike and dip.

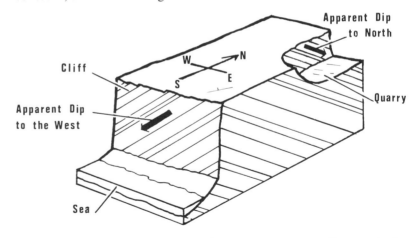

Figure 12.3 Apparent and true dip. There are two apparent dips seen in the flat quarry wall (to the north) and the cliff-face (to the west). The three-dimensional pattern is necessary to find the true dip, which is to the north-west.

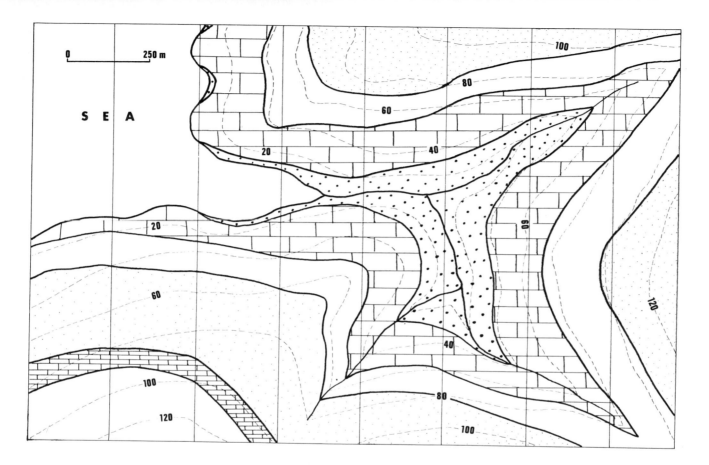

Figure 12.4 Tilted rocks. The strike lines are drawn across the map so that you can see how to place them. Calculate the dip gradient and direction. NB Strike lines on geological maps must always pass through an intersection between a relief contour and bedding plane outcrop. Heights are given in metres.

Dip and Strike. Make a rough calculation of the dip of these rocks. The photograph was taken at Croyde Bay, north Devon looking westwards out to sea. In which direction is the strike? (*Crown Copyright*)

The dip of strata affects the patterns of rocks as they reach the surface—ie in their outcrop. If a group of rocks are in the horizontal position their outcrops will run parallel to the contours, whereas if the dip is vertical (90 degrees) the rock will form a narrow, straight outcrop across hills and valleys alike, resembling a wall. When the rocks are tilted at an angle between these extremes, the pattern of outcrops in the valley will depend on the angle of dip and the slope of the valley floor: Figure 12.4 shows some of these relationships.

3 What patterns of folding or fracturing affect the rocks? As we find out more and more about the rocks in an area we shall see patterns emerging in the structures which affect them. The most important are folds, when the rocks are bent, and faults, when they are broken, but there are also minor breaks in the rocks without displacement on either side, and these are known as joints. Folding and faulting are caused by pressures being applied to rocks. Such pressures sometimes alter the nature of the rock (Chapter 11), but always cause some degree of deformation. If the affected rock is still soft and relatively flexible, it will bend into continuous folds, rather like the ones you can make by crumpling the tea-cloth on a table. If it is brittle, however, the rock will break as stress is applied. Thicker masses of sediment are folded more easily than thin sediments lying on a rigid basement of rocks, as is demonstrated in southern England, where the Jurassic and Cretaceous deposits are many thousands of metres thick. When the Alpine folding movements took place they were compressed into open folds, whilst the thinner rocks of the same age north of the river Thames were only tilted and fractured. Occasionally a very weak bed of rock will collapse completely and flow (Figure 8.14): we have seen how 'plastic flow' occurs in substances like ice and salt, but it can affect a wide range of rocks if a stress is applied for long enough.

Folds

There are two types of folds, illustrated in Figure 12.5. The upfolds are **anticlines**; the downfolds are **synclines**. These basic patterns may vary in a number of ways (Figure 12.6): folds may be symmetrical,

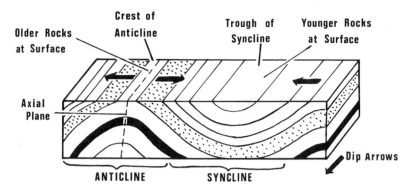

Figure 12.5 Simple folds. Notice that the rocks at the surface get younger towards the centre (trough) of the syncline, and older towards the centre (crest) of the anticline.

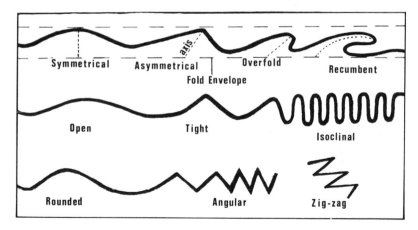

Figure 12.6 Fold patterns. A few of the patterns formed during the folding of the rocks.

185

Asymmetrical folds in an opencast coalmine, Carmarthenshire. (*Crown Copyright*)

with vertical axial planes, or asymmetrical when the axial plane is inclined to the fold envelope. The term **monocline** is sometimes used for a fold with one very steep limb between horizontal, or nearly horizontal, limbs.

Some are overturned to form recumbent folds, in which the strata on one limb are inverted. They may be open, or tightly pressed together, and they may be rounded or angular in cross-section. All these characteristics are determined by the intensity of the folding pressures, and the softness and flexibility of the rocks when they were folded. A fold may also dip at one or both ends—when it is said to plunge.

The bending of rocks as they are folded often means that the top of an anticline begins to crack open, just as a rubber does when you bend it in half. The joints caused in this way may be filled by percolating solutions, or may form a zone of weakness on the fold crest which is easily attacked by weathering and river action. Although the anticlines form the hills at first, and the synclines the valleys, it is common for this arrangement to be reversed after a prolonged period of erosion. The Vale of Pewsey to the north of the Salisbury Plain is an example of such **inverted relief** since it is a low-lying area along the line of an anticline.

Some of the most spectacular results of folding are to be seen where alternating bands of massive, resistant rock and weak shale are involved. The resistant layers are known as **competent**, and the weak as **incompetent**. Whereas the competent beds resist the stresses affecting the rocks, and form rounded folds, the incompetent rocks crumple easily and are strongly distorted. Some of the results, including overfolds and zigzag folds, can be seen clearly in the cliff rocks of Upper Carboniferous age around Broad Haven, west Pembrokeshire.

The most intense folds are only found in certain regions, and are usually on too large a scale to be traced in a quarry wall, or cliff-face. The Scottish Highlands and the great ranges of fold-mountains such as the Alps, Himalayas and Andes are formed of large-scale folds of the isoclinal and overfold types as well as vast broken slices of rock.

Intensely folded regions may also have experienced metamorphism, which imparts a cleavage to the rocks. This type of structure enables

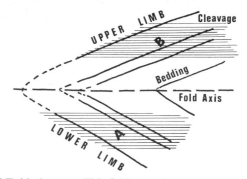

Figure 12.7 Fold cleavage. This fold was observed and sketched at Bovisand Bay, near Plymouth. Both limbs of the fold are dipping to the south, and this means that the rocks of the lower limb are in reverse order (ie the oldest is on top if you draw a vertical line through them). Devise your own rule to determine which way up the rocks are, using the information provided by the dips of the bedding planes and cleavages.

the geologist to determine which limb is the correct way up in overturned rocks (Figure 12.7).

Faults and Thrusts

Faults are fractures of the rocks in the Earth's crust where there is displacement of one side relative to the other. They may be caused when rocks are crushed together, or when they are pulled apart—by compression or tension as shown in Figure 12.8. The latter seems to be more common, since **normal faults** are formed in this way, whilst **reverse faults** and **thrusts** originate when one group of rocks are pushed over another, and a section of the Earth's crust is shortened. Both involve vertical movements in the rocks, but **tear or wrench faults** affect them by sideways movements. Faults may also be described by reference to their relationship to the directions of dip and strike in the rocks affected. Thus a fault with an outcrop parallel to the strike of the rocks it fractures is known as a **strike fault**; one parallel to

A simple anticline, cut open by erosion in Iran. It plunges towards the top left-hand corner. Notice the part river erosion has played in destroying the rocks in this arid region. Make your own sketch-map of this situation, numbering the layers of rock in the order in which they were formed. (*Aerofilms*)

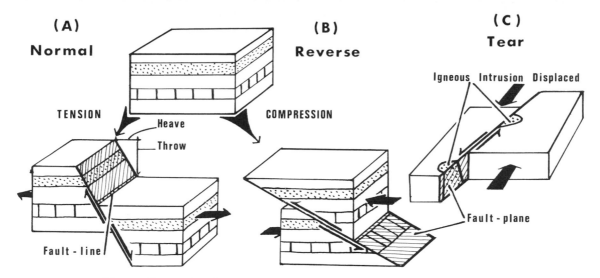

Figure 12.8 Faults. Some faults involve mainly vertical movements up or down the dip of the fault plane (normal and reverse), and some mainly horizontal movement along the strike of the fault plane (tear).

the dip as a **dip fault**; and one between these directions as **oblique**. (NB Another set of terms for fault patterns is also used. Dip-slip faults are those involving a movement down the dip of the fault plane—ie what we have termed normal and reverse faults with a largely vertical movement. Strike-slip faults involve movements along the direction of strike on the fault plane, or horizontally: they include the tear fault.)

When we examine a small fault closely we often find that there is a zone of fracture a few centimetres wide which is occupied by a mass of small angular fragments and fine clay, produced during the fault movement. If this fault breccia is not present it is common to find the fault planes grooved and scratched by the movement: this feature is known as **slickensides**. In many cases larger faults, extending over hundreds of kilometres, are zones of several near-parallel fractures rather than a

single line of breakage. Such faults often do not have a clear outcrop: as zones of weakness they may be occupied by a gully and covered with weathered debris. The outcrop patterns of the rocks on either side of this gully will provide the evidence for the fault's existence, as Figure 12.11 shows.

Normal faults are often found in two parallel groups where large masses of rock have been let down between them to form **rift valleys** like those of the middle Rhine (illustrated in Figure 12.13) and East Africa. Both have been caused by local uplift of an elongated dome and the collapse of the central part to form a long trough over 40 km across and bounded by a stepped series of normal faults. Lake Victoria is formed in the wide depression between the two updomed areas which have been rifted in East Africa.

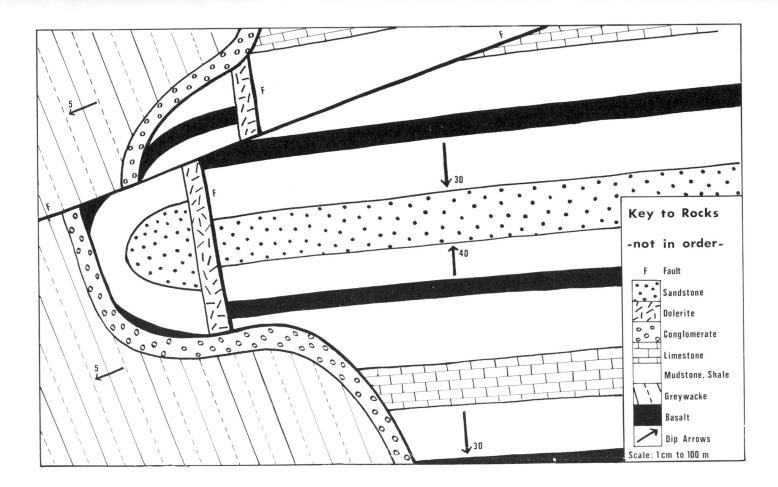

Key to Rocks

-not in order-

F   Fault

Sandstone

Dolerite

Conglomerate

Limestone

Mudstone, Shale

Greywacke

Basalt

Dip Arrows

Scale: 1cm to 100 m

Figure 12.9 Map exercise. What evidence does the map record which suggests that the rocks in this area have been folded? Write out a list of the geological events which have taken place.

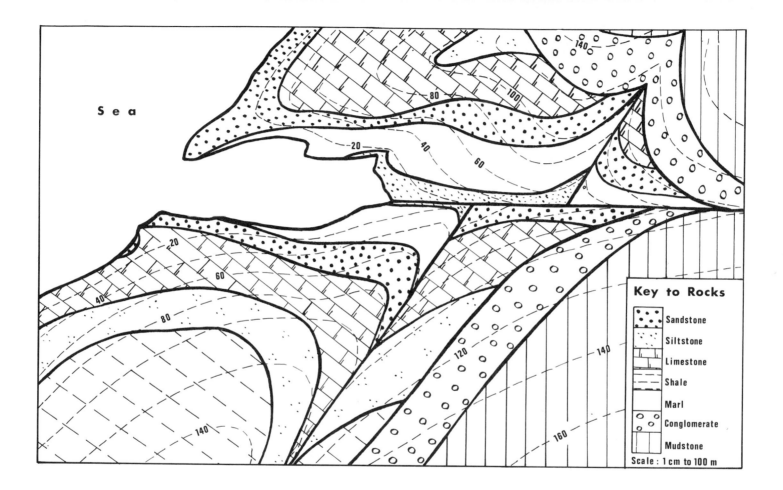

Figure 12.10 Map exercise. Give an account of the faulting (type and date) which has affected the rocks of this area.

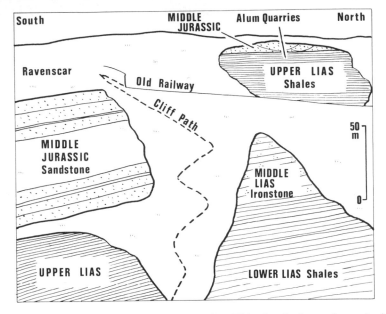

Figure 12.11 Robin Hood's Bay, Yorkshire. This sketch shows the rocks in the cliffs at the southern end of the bay. Draw your own diagram and place in a vertical fault line. Write down the evidence on which you base this position and calculate the throw of the fault.

The opposite effect, whereby a faulted mass is uplifted to form a **block mountain, or horst**, is also common. Areas like the Harz Mountains of Germany, and the Longmynd area in Shropshire, have been raised in this way (Figure 12.13).

Reverse faults are most commonly found as low-angle thrusts, especially in highly folded areas where the degree of deformation is greater than that to which the rocks can adjust in simple folds. Thrusts often replace the inverted limbs of recumbent folds (Figure 12.12). In the Alps there are piles of thrust slices of rock known as **nappes**, which

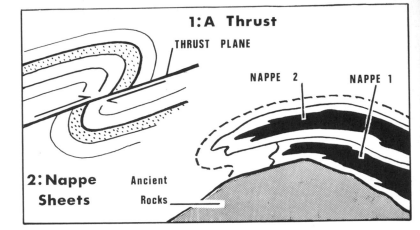

Figure 12.12 Thrusts and nappes. Two structures resulting from extreme deformation of the rocks.

have been carried over the top of younger rocks and unaffected resistant basement rocks. An important thrust zone is found in the Highlands of Scotland, where the folded and metamorphosed Precambrian Moine Series of rocks are thrust over unmetamorphosed Torridonian and Cambrian strata. The section in Figure 18.2 illustrates the results, and shows the extent of the broken zone of dislocated rocks often altered to mylonite. It has been shown that some of these rocks have travelled at least 16 km from where they were originally deposited.

Tear faults involve horizontal movements on either side of the fault plane. The Great Glen fault in northern Scotland is a broad belt of shattered rocks where there has been a sideways movement of 100 km as indicated by the displacement of a granite on either side of the fault. This is similar to the effect shown in Figure 12.8C; compare the positions of the Foyers and Strontian granites on a geological map. The northern part of Scotland slid 100 km to the south-west, mostly

A Tear Fault: the Great Glen in Scotland. There is evidence to show that the northern side (on the left) moved 100 km to the south-west. (*Aerofilms*)

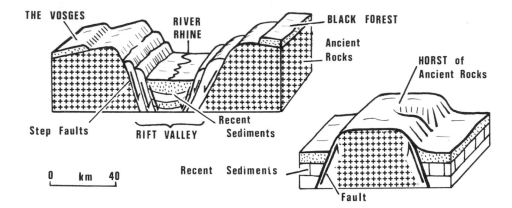

Figure 12.13 Rift valleys and horsts. Rift valleys are formed as a section of the earth's crust is let down by the collapse of the centre of a mass of updomed rocks. Horsts often result from the effect of earth movements on hard, ancient rocks.

during the late Caledonian movements at the end of the Devonian period, but minor earthquakes are still common in the area. The best-known tear fault is the San Andreas fault in California (Figure 12.14), which is 1300 km long and has nearly 2 km of brecciated rock in the fault zone. It has been moving since the late Jurassic, and the Pacific side has been moving northwards, wrenching open the Gulf of California and causing some areas to move over 500 km in that time. The movement has taken place in small stages, including the disastrous San Francisco earthquakes of 1906, when displacements of a few metres were enough to break gas mains and cause many fires. The average movement of this fault is to separate the land on either side by 4 km every million years, but at present, in an era of more intensive earth activity the average rate is 50 km per million years.

This fault zone tends to move in sudden jerks after a period of years when stresses build up. San Francisco today once again stands in danger of a major calamity, although its inhabitants seem to be extremely casual about it: massive office blocks, schools, hospitals and even the disaster headquarters, have all been built across the line of the San Andreas fault itself, where the greatest damage would occur. Observations in the 1960s have suggested that a fault which is lubricated by pumping a liquid into it moves in more frequent, shorter stages: it will give earthquakes more often, but their effect is very slight. American scientists are now trying to apply this treatment to the San Andreas fault, so that they might avert the impending disaster.

Folds, Faults and Time

How can we tell the age of a fold or fault? We usually have at least two sets of criteria. The folding or faulting movements took place **after** the youngest rocks they affect, but **before** the oldest undisturbed sediments lying on top. The Carboniferous rocks of the Pennines are folded and

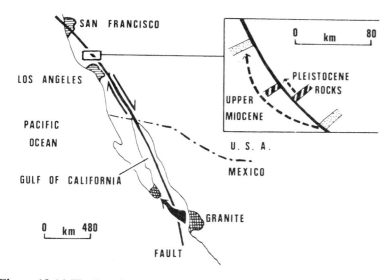

Figure 12.14 The San Andreas Fault. The enlarged inset shows some detail of the movements. Younger rocks of Pleistocene age (1 million years old) have been separated by 18 km only in the course of successive movements along the fault; those of Miocene age (15 million years old) have been separated by 100 km. At the entrance to the Gulf of California a mass of granite has been moved 500 km apart.

faulted, but the overlying Permian rocks are not affected in the same way. We date the movements which caused the Pennine folding and faulting as late Carboniferous because the uppermost Carboniferous rocks found elsewhere are missing. An unconformity usually marks the division between the older folded or faulted group of rocks and the overlying unaffected younger series (Figure 17.9: compare this situation with Figure 12.16).

Faults may have long histories of recurrent activity, as in the case of the San Andreas fault already mentioned. The Highland Boundary fault may have been initiated as early as the Ordovician and still causes slight earthquakes, eg at Comrie in Perthshire. Such faults appear to mark major lines of crustal weakness.

Joints

Many cracks we see in rocks are not faults because the rocks are not displaced on either side. These partings are joints, and are due to stresses of different sorts from the earth movements we have just been discussing. They are important in affecting the detailed shape of

Jointing in the Giant's Causeway, Northern Ireland. The fine-grained rock contracted around a number of centres within the lava, causing the cracks to form. Note the variation in the number of sides between the columns. (*Crown Copyright*)

landforms and drainage patterns, since they are lines of weakness and easy entry for weathering agents. There are several types of joint.

(1) Sedimentary rocks are criss-crossed by **shrinkage joints**, formed as the drying rock became more compact.

(2) Vertical **cooling joints** are common in lavas, where the shrinkage took place for another reason. The hexagonal columns of the Giant's Causeway and the Isle of Staffa are famous scenic wonders. The columns develop in directions perpendicular to the cool surfaces over which lavas flow, and they may be curved.

(3) Igneous intrusions may be crossed by **parallel joints** which cause great sheets of rock to come loose. The sides of the Goat Fell granite in the Isle of Arran are characterised by such 'boiler plate' formations, which are thought to have been caused by the expansion of the rock as the original pressures were released by the removal of the overlying rocks. These intrusions also have groups of joints at right angles to each other, formed during cooling, and these often affect the drainage pattern, as in the case of Bodmin Moor (Figure 5.6C). Joints connected with igneous rocks are liable to be filled with pegmatite, or mineralised veins as the fluids and gases continue to circulate through the rocks at a late stage in the cooling process.

(4) Other joints are the direct result of the stresses occurring in earth movements, particularly where folding is involved, and are illustrated in Figure 12.15.

The Formation of Folded Mountains

So far we have mentioned two ways in which land may be raised up to form mountainous masses: we have studied the formation of volcanic cones, and of horsts. The most important and highest mountains in the world are all folded mountains—chains of very high land stretching for thousands of kilometres. The Andes, Rockies, Alps and Himalayas are all 'young' folded mountains. They were raised up in relatively recent time during the Tertiary period, and the Himalayas are the highest largely because they were the last group to be formed. Erosive forces have cut deeply into the Himalayas (Figure 5.13), but have

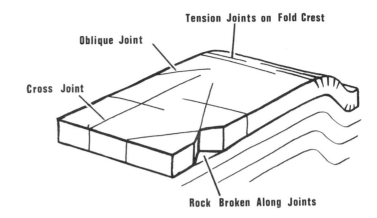

Figure 12.15 Joints. This diagram shows where lines of weakness form in a rock due to jointing following folding.

scarcely lowered the summits, whereas they have already reduced the Alps to half of what were probably their greatest heights! Older groups of folded mountains have been completely worn down to peneplains in the past: these can be seen when we examine unconformities, and on the tops of block mountains, where a large mass of peneplained land has often been uplifted between two faults.

Some parts of the Earth are subjected to repeated Earth movements, including earthquakes and volcanic action (Figure 10.1), whilst others experience few changes. The mobile belts contrast with the stable areas in many other ways. The mobile belts are long, narrow, sinuous zones in which great thicknesses of sediment accumulate, whilst the stable areas of the Earth's crust cover much larger expanses and have relatively thin layers of sediment forming on top. The mobile belts are associated with large-scale **orogenic**, or mountain-building events, whilst the stable areas are affected only by **epeirogenic** warping and faulting.

The mobile parts of the world at the present time are therefore those

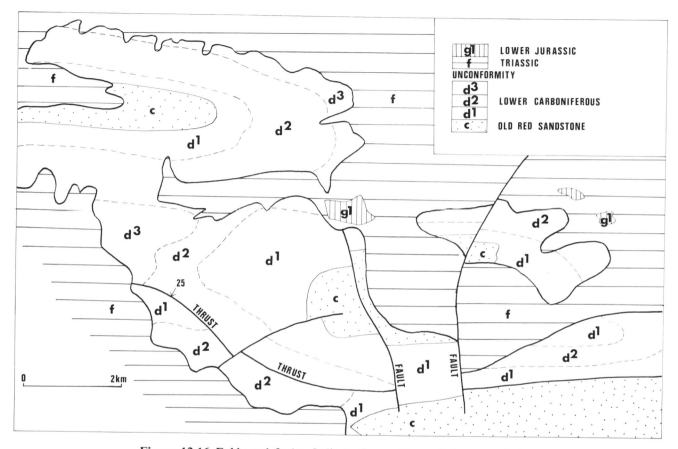

Figure 12.16 Folds and faults. Indicate the positions of the axes of the anticlines and synclines on a sketch-map of this area. Draw in a major unconformity. Describe the nature of at least two different types of fault. Suggest dates for the formation of the folds and faults. This is a sketch-map of the geology of part of the area near Wells in Somerset, and is based on the maps of the Institute of Geological Sciences. (*Crown Copyright Reserved*)

areas where earthquakes and volcanoes are most common. The deepest ocean trenches are all situated close to these belts, occurring off the coasts of the East and West Indies, the Philippines and Japan. One of these, the Tonga Trench, is 1100 km long, 60 km across and over 9 km deep, whilst the Mohorovičić Discontinuity is 9 km deeper than usual beneath it. These trenches lie parallel to lines of volcanic islands, like the Aleutians, which often pass into ridges of folded mountains.

It is thought that the deep trenches are similar to the early stages in the development of some of the great troughs of sedimentation which were responsible for the piles of rock involved in the formation of folded mountains in the past. Other troughs seem to have been filled with shallow water sediments: the rate of sedimentation must have balanced the rate of deepening. These troughs are known as **geosynclines**, and it is in them that the story of folded mountain-building has always begun. This much is clear, but when we seek the explanations for the series of events we are about to describe in detail we have to come to the conclusion that we do not yet know the answer. It is realised that geosynclines and folded mountains are the result of some complex movements inside the Earth, and in the past they have been related to the contraction of the Earth, and to the idea that our continents have been moving about on the Earth's surface. In recent years the idea that there are very slow convection currents circulating in the Mantle has gained favour. These are beneath the crust, and tend to drag down sections of the surface rocks where the currents turn downwards, forming a long zone of subsidence.

We can trace the series of events from the rock record, and the diagrams of Figure 12.17 will help us.

1 **The Geosynclinal Phase** (Figure 12.17(1)). A geosyncline is formed when a long, narrow section of the Earth's crust becomes weak and sags downwards, forming a deep ocean trench or an area of shallower seas with long-continued subsidence. At first in the deeper waters the sediments may accumulate slowly as fine mud, containing a high proportion of undecomposed organic matter in the deep, stagnant waters. The surrounding lands tend to rise and sooner or later great quantities of sediment are poured into the geosynclinal sea. Volcanic action is common, and thick layers of spilite lava and tuff form. Earth movements affect the surrounding land masses to an increasing extent, and the intensity of erosion works up to a climax when thick piles of greywackes build up. The trench is eventually filled in with up to 10 km of sediment and lava. It is a very gradual process, and usually takes anything from 100 to 200 million years.

2 **The Early Orogenic Phase** (Figure 12.17(2)). Finally the geosynclinal sag ceases to subside, and the sediments which have filled it begin to be compressed into folds, thrusts and nappes. All this begins during the later stages of the geosynclinal phase when there are a series of short-lived Earth movements separated by quieter periods of continued sedimentation (a series of alternating shales and coarse greywackes known as the **flysch facies**). Small-scale, local unconformities are a feature of the rocks formed at this time. The most deeply buried rocks become metamorphosed under conditions of great heat and pressure, and they are often granitised or impregnated partially by outside fluids to form migmatites. At this stage the mobile geosyncline lies between two areas of resistant, foreland rocks, and as they move together like the jaws of a vice the rocks in the middle are compressed.

3 **The Late Orogenic Phase** (Figure 12.17(3)) is the mountain-building stage. The folded and metamorphosed sediments and lavas are uplifted into the towering chains of folded mountains with which we are familiar. The whole process is probably caused to a large extent by isostatic adjustment and is accompanied by powerful thrusting movements as the soft rocks are squeezed upwards. As soon as the mountain ranges are formed erosion begins, and marginal troughs and basins are rapidly filled with arkose-type deposits (ie **molasse**). The intrusion of granite from below takes place during this late stage. The whole orogenic phase also lasts a long time, partly overlapping the period in which the rocks accumulated: the deformational movements

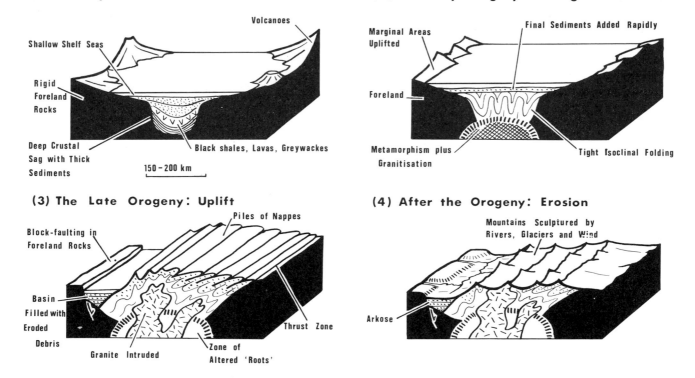

(1) A Geosyncline: the rocks

Volcanoes

Shallow Shelf Seas

Rigid
Foreland
Rocks

Deep Crustal
Sag with Thick
Sediments

Black shales, Lavas, Greywackes

150 – 200 km

(2) The Early Orogeny: folding

Marginal Areas
Uplifted

Final Sediments Added Rapidly

Foreland

Metamorphism plus
Granitisation

Tight Isoclinal Folding

(3) The Late Orogeny: Uplift

Piles of Nappes

Block-faulting in
Foreland Rocks

Basin
Filled with
Eroded
Debris

Granite Intruded

Zone of
Altered 'Roots'

Thrust Zone

(4) After the Orogeny: Erosion

Mountains Sculptured by
Rivers, Glaciers and Wind

Arkose

Figure 12.17 The formation of folded mountains.

during the Alpine orogeny lasted from the late Cretaceous to the Miocene period (ie up to 60 million years), but rocks were still being laid down and earthquakes in these areas today testify to continued, but waning, activity. The axis of a geosyncline may migrate so that folding in one part is paralleled by subsidence and sedimentation in another.

4 **The Post-Orogenic Phase** (Figure 12.17(4)). As the mountain

ranges become more stable the igneous activity and erosion begin to slow down, but it is a long time before a mountainous area becomes 'quiet'. During this stage there is an unusually large proportion of dry land, and continental deposits including desert sands and evaporated salts are characteristic. Eventually the area is completely worn down and shallow seas transgress the exposed mountain 'roots'.

A Pattern of Earth Movements

When we study the geology of an area like Europe, or North America, we find that the history of these areas can be summarised in terms of a series of orogenies. Figure 12.18 shows the pattern in Europe, which is so important as we come to a consideration of the events leading to the building up of the British Isles. During the Lower Palaeozoic (Cambrian, Ordovician, Silurian) a geosyncline extended from Ireland to northern Norway and dominated central and northern Britain. The mountains formed from the rocks of this geosyncline at the end of the Silurian were eroded, and much of the debris was transported into an east–west geosyncline across central Europe in the Upper Palaeozoic (Devonian, Carboniferous). Southern Europe then became the site of a third geosyncline which extended eastwards to the position of the present Himalayan Mountains. This Tethys geosyncline lasted from the Permian, through the Mesozoic (Triassic, Jurassic and Cretaceous) and into the Tertiary, when further folding movements took place and the Alps were formed. We live in the dying-down phases of this orogeny, and erosion is wearing down the mountain ranges.

This almost makes it look as if Europe has grown by the addition of new ranges of mountains around the original 'core' of the Precambrian Scandinavian Shield. In fact the foundation rocks of each successive geosyncline have been those which were involved in similar events at an earlier date: rocks folded in the late Silurian orogeny include slices of the underlying Precambrian basement; the Alps contain large masses of Hercynian and Precambrian granites within the great mass of folded sediments. Europe has not grown noticeably, but the centres of weakness and geosyncline formation have shifted.

So much is clear when we study the rocks of Europe, and we must bear these facts in mind as we study the historical geology of Great Britain in the next section. Recent investigations of Precambrian rocks show that the same held true, and a series of earlier orogenies have been recognised in the Scandinavian 'core' of the continent (Figure 12.18).

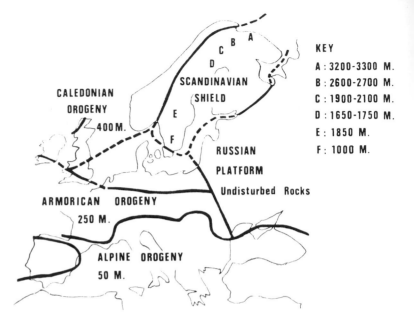

Figure 12.18 European orogenies. Note the many orogenies now recognised in the Scandinavian Shield area of Precambrian rocks. The numbers (eg 400 M) are the ages of the orogenic movements in millions of years.

Geology and Engineering

Civil engineers employ a knowledge of geology in their work, whether it be in the study of the foundations of buildings, roads, dams or bridges they are helping to erect, or the material into which they are tunnelling. They are particularly interested in such factors as the load-bearing strength of the rock, its permeability, and whether it has been weakened by folding, faulting or jointing. Mining engineers must take account of the frequency of faulting and the dip of the rocks before they suggest the investment of large sums of money in the sinking of shafts and the purchase of mining equipment. In spite of the normally adequate preliminary surveys at Bevercotes Colliery in Nottinghamshire, a higher density of faults has been discovered in practice than was expected, and this has made it difficult to use the expensive, automated machinery which had been installed.

Extensive geological surveys are necessary before such structures as the Severn Bridge can be designed and built. When the Burrator Dam, north-east of Plymouth, was being constructed it was discovered that a large area of granite rock had been rotted by weathering: this was not permeable, but quartz veins crossing it had become broken and were extremely permeable. The whole mass of rotted rock had to be removed and an extra length of dam built.

The London Clay formation beneath London is festooned with the tunnels belonging to the underground railway and the major sewers: it is an almost ideal tunnelling medium, allowing rapid progress for the excavators. When the Victoria Line was being built in the 1960s it was sometimes necessary to go upwards into the Thames river gravels, or downwards into the sands and gravels of the Woolwich and Reading Beds in order to avoid older tunnels. This necessitated costly and slower progress in order to prevent water entering or the tunnels collapsing.

THINGS TO DO AND DISCUSS

Investigation 1. Rock structures

Examine a quarry face for all the evidence of distinctive structures in a layer of rock and along its bounding bedding planes.

(1) Record the strike and dip of the bedding planes and any fault planes; the direction and spacing of the joints, the bedding plane features and the internal rock structures. Draw a labelled sketch of your findings.

(2) Divide the structures you have recorded into (a) those which originated in the processes of rock formation, and (b) those which are tectonic in origin.

(3) Write an account of what you can discover about the history of the rock from this study of the internal structures.

Investigation 2. Engineering geology

Find out the details about a mining or engineering situation, preferably near your home or school, in which geological knowledge has been important.

Making models of geological situations

It is often a valuable exercise to make a relief model of a particular area whose geology is known to you and to paint on it the rock outcrops. This will help you to understand the relationship between geological maps and geological sections.

(1) Simple blocks of polystyrene can be used to demonstrate some of the outcrop patterns as they occur on maps. Use felt-tipped markers to draw on the geological structure. Then carve V-shaped valleys into the top surface and redraw the outcrop pattern. Begin with tilted rocks and notice the patterns made by valleys cutting into them; then progress to symmetrical and asymmetrical folds; add an unconformity; and work up to faults of various types. Summarise your work by drawing a simple sketch-map of the situation and the section on the side of your block.

(2) *More sophisticated models* can be made of actual geological situations, building up the relief from an Ordnance Survey map, and adding the coloured geological outcrops.

Materials required. A baseboard which will not bend easily (eg $\frac{1}{2}-\frac{3}{4}$-inch chipboard), cut to size; Polyfilla; polystyrene ceiling tiles; a battery-operated, hot wire cutter for the tiles; adhesive for tiles; tracing paper; modelling tools; basic colours in powder paints; varnish; turps for thinning; transfer lettering (eg Letraset). Many of these items can be obtained in the school, or cheaply outside.

Method. Determine the scale of map to be used as a basis, and relate this to the $\frac{1}{4}$-inch thickness of the ceiling tiles for a vertical scale (eg tile = 400 feet on a map scale of 4 miles : 1 inch; 200 feet on 1 mile : 1 inch; 50 feet on 1 mile : 6 inches; these figures will need to be metricated when tiles and maps come in metric sizes and scales). Trace out the relevant contours on a large sheet of tracing paper, and transfer them, with the trace of the coastline, to the baseboard. Having done this cut out and pin sections of the tracing paper on the ceiling tiles, and cut round the traced contours with the hot wire cutter; stick each section to the baseboard, and stick higher contour areas on the lower. Then with the map in front of you use a thick mixture of Polyfilla to round off the slopes and eliminate the stepped effect; go over each completed section with water to smooth the surface. Allow to dry. Mark on the areas of rock outcrop. For each area make up sufficient powder paint colour in a thick paste to cover the entire area and the space in the key drawn at the side of the model. When the painting is finished allow to dry. Paint the whole model with a thinned varnish. Allow to dry. Letter salient points, the key, a title. Finally add another coat of full strength varnish.

Questions for discussion and essay-writing
(*a*) Explain, with the help of diagrams, the manner in which the folding and faulting of solid rocks may influence the scenery, giving actual examples where you can (S).
(*b*) Explain what you understand by the geological terms, true dip, apparent dip and strike. Wherever possible, illustrate your answer by referring to the geology of an area known to you (O).
(*c*) Describe minor structures which indicate whether sedimentary rocks are the right way up (O).
(*d*) With the aid of diagrams explain the difference in each of the following pairs between: (i) an outlier and an inlier, (ii) a normal fault and a reversed fault, (iii) a monocline and a syncline (N).

(*e*) Describe what is meant by the term 'mountain building'. Quote examples from the main episodes that affected Great Britain (AEB).
(*f*) Distinguish between a joint and a fault. Describe how joints and faults may influence topography (L).
(*g*) Draw block diagrams to show: (i) a plunging asymmetrical syncline, (ii) a reversed fault concealing the outcrop of certain strata and (iii) an unconformity with horizontal rocks overlapping on to folded strata (L).
(*h*) Explain simply why folds, faults and unconformities each indicate movement in the Earth's crust. Draw labelled diagrams to illustrate actual examples of two of them (W).
(*i*) Describe the principal geological structures which permit the accumulation of a water supply (O).

Folded and faulted rocks in arid Iran. (*Aerofilms*) ▶

Part Four

Earth Patterns

13 The Earth's Interior and Origin

Earth Patterns: Introduction

The study of the Earth's relief features, of the geological processes acting on the surface rock and of the surface rocks themselves, has provided us with much evidence, which can be used to work out the past history of this planet or in the search for new resources. Certain patterns emerge from such studies, which enable us to reconstruct the past history of the Earth. These are the concern of this part of the book.

We shall first study the evidence relating to the constitution of the Earth's interior, and this will lead on to a consideration of the possible origin of the planet and the major layers of which it is formed. Then we shall pass on to a consideration of the evolution of the main surface features—the continents, ocean basins and mountain ranges. All these patterns are based on world-scale evidence and involve the major relief entities and geological activity taking place at the Earth's surface. The final part of this section brings together the more local evidence, which can be used to work out the detailed history of a smaller area. We shall study the principles on which such a study is based, and then apply these to the British Isles, within whose borders resides one of the most varied geological situations in the world. It was here also that much of the early work in modern geology was carried out.

It must be remembered, however, that at this stage we are moving from the established, observed facts, which can be measured and verified again and again, to the realm of interpretation and even speculation. New facts come to light and interpretations change. Each fact upon which an idea is based must be looked at critically, especially when a principle which has been found to work with one set of facts is extended to another.

It is worth recording that the concept of continental drift, which is now receiving so much attention and acclaim, was derided by most geologists in the mid-1950s: new facts have emerged, we are told, and this has altered the situation. It is a healthy sign, however, if we question this new evidence and see if it is as indisputable as we are often told. How far, for instance, is the concept of ocean-floor spreading backed up by direct observation, and how far is it based on the unwarranted extension of certain trends?

And yet, whilst we must question the basis for these interpretative patterns, they have a value, albeit often temporary, in pointing the way forward for further investigations. They provide us with an overall view of the world and of the geological activity taking place at its surface. Many questions still remain unanswered, and it is the nature of the subject that they will probably remain so, but the modern interpretation of Earth phenomena is a stimulating and satisfying field of study.

The Earth's Interior

Our knowledge of the Earth's interior is of a different order to that of the surface rocks. No one has been more than a few kilometres down into the Earth, and we must use a number of indirect methods of observation and infer the possible nature of the interior from the results gained.

Men have studied volcanoes spewing out molten lava at extremely high temperatures, and have noticed that the temperature rises as they descend very deep mines. Direct evidence of this type has led to the conclusion that the interior of the Earth is much hotter than the surface: temperatures may be as much as 500°C 30 km beneath the surface, and 6000 degrees in the deep interior. We must remember, however, that pressure also increases towards the Earth's centre because of the weight of the overlying rocks. This raises the melting points of the rocks, and one which melts at 800°C at the surface may not do so until several thousands of degrees in the deepest parts of the planet.

Most of our knowledge of the Earth's interior has been obtained by indirect methods, and the study of earthquake records has been of particular value. We have all seen pictures of the terrible results of earthquake movements, and recently there has been a number of violent tremors in such places as Chile, Alaska, Turkey and Yugoslavia. The main zones of earthquake activity are plotted in Figure 10.1. These shocks have been caused by pressures acting on the rocks down to as much as 720 km beneath the Earth's surface, though most of them occur at depths of 60 km or less, the deepest only happening in the Pacific Ocean area. When the rocks give and fracture, or move along old fracture lines known as **faults**, great stresses are produced and shock waves are transmitted through the Earth. Some of these waves travel round the outside of the planet, causing damage to property and upheaval of the surface; others go straight through.

The intensity of earthquake shock waves can be recorded on an instrument known as a **scismograph**, and there are many of these instruments taking continuous recordings throughout the world. The principle used in these seismographs is illustrated in Figure 13.1, and when an earthquake takes place a record like that shown in Figure

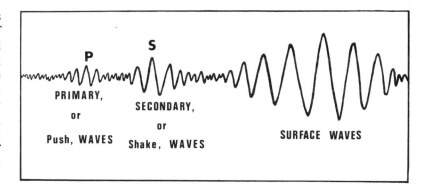

Figure 13.2 A seismograph record. The recording moves from left to right. The P and S waves arrive first because they travel directly through the Earth; the surface waves make the most powerful impression, but pass only along the surface of the planet.

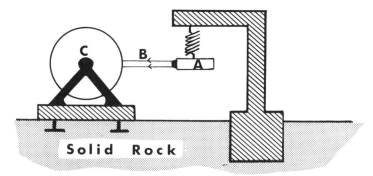

Figure 13.1 A seismograph. A motionless pendulum—A—is suspended from a metal frame anchored in the rock. A beam of light—B—is projected on a rotating drum—C—covered with photographic paper. The drum is anchored in the rock and vibrates when an earthquakes occurs: this leaves a wave-like record on the photographic paper.

13.2 is produced with three successive sets of shock waves. The actual position of the earthquake can be plotted when the records of at least three seismograph stations have been compared. Seismographs nearest to the centre of the earthquake will record the most intense shocks, and the **epicentre** is the point on the surface where the movements have been greatest. The earthquake **focus** is the point at depth where the actual movement took place.

We can learn still more. Each of the different types of shock waves recorded (Figure 13.2) has different characteristics, though the P and S waves are most important because they travel through the Earth. The P waves travel fastest, and therefore arrive at the seismograph station first. Whereas the P waves are compressional, and the particles in their path vibrate back and forth, the S waves are transverse with side-to-side movements, and will not pass through liquids. Figure 13.3 shows what four seismographs distributed around the world record when an earthquake takes place. Stations A and B receive all the shock

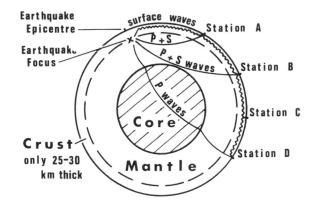

Figure 13.3 The Earth's interior. The use of earthquake shock waves to work out the nature of the Earth's interior. A, B, C and D are four seismograph stations recording shocks from the earthquake at X. Note the type of wave each station receives, remembering the characteristics of the three groups.

Seismic waves are deflected (refracted) and their velocities changed when there is a change of density in the rocks through which they are passing. Seismograph records are usually more complicated than the one we have considered so far, with two or three sets of P and S waves arriving at the receiving station—indicating a layered structure above the Earth's core. The most pronounced change in the velocity of the seismic waves occurs between the rocks of the thin Earth's **Crust** (with densities between 2·7 and 3·0 gm/cm³) and the underlying **Mantle** (density 3·3 gm/cm³). The boundary between the two is known as the Mohorovičić Discontinuity (or just 'Moho'), after the Yugoslav seismologist who discovered it in 1909. A study of Moho tells us that the Crust is thickest under the continents and very thin under the oceans. One of the more minor divisions shown up by earthquake records is one within the Crust. There is a lower zone, similar in average composition to the rock basalt and often known as the **sima** (it is composed largely of silica and magnesia) or **basaltic crust**, and an upper zone of lighter rocks similar to granite known as the **sial** (silica and alumina) or **granitic crust**. These divisions are illustrated in Figure 13.4. The Earth is thus composed of a series of shells getting lighter and lighter from the centre to the surface.

The Crust is thickest beneath the continents, and especially beneath the mountain ranges. This discovery was first made by surveyors who mapped differences in gravitational pull over the Earth's surface. It is to be expected that high mountain ranges like the Himalayas will attract a plumb-line because of their extra mass. Scientists were puzzled when they found that the deflection of the plumb-line towards the mountains was only one-third of the value they had calculated. They realised there were two possible explanations. Either the mountains were formed of much lighter rocks than the surrounding areas, or the lighter surface rocks extended to greater depths beneath the mountains. It was soon shown that the surface rocks do not vary in this way, and the latter theory has been adopted. There are deep 'roots' under each range of mountains.

We would notice a similar result if we examined two icebergs of different sizes. Icebergs float on water because they are less dense, but

waves first with the most powerful effects, and they have records similar to Figure 13.2. Station C only receives surface waves, since the S waves have been obstructed and the P waves slowed down and refracted by a different material in the Earth's centre. Station D receives P and surface waves, but with weakened signals. From such records we can conclude that there is a central mass inside the Earth, known as the **Core**, which will not allow S waves to pass through, and which slows down the P waves. It is very possible that this is liquid. Sir Isaac Newton's Law of Gravitation has helped us to calculate the mass of the Earth. Since we also know its volume, we can work out the average density to be 5·527 gm/cm³. But we know that the surface rocks have a density of less than 3, and have to conclude that the rocks of the Core must be much denser—about 12: they are probably a mixture of nickel and iron like many of the meteorites which have been found.

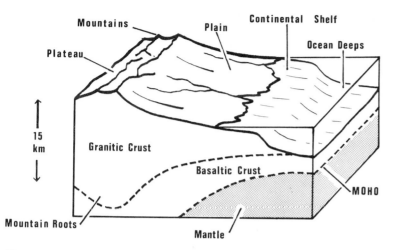

Figure 13.4 The Earth's crust. Note the relationship between the surface relief and the thickness of the crust.

Figure 13.5 Isostasy (1) Icebergs. Two icebergs, A and B, have a similar proportion of their mass out of the water, even when some ice is transferred from one berg to the other—the state of equilibrium between ice and water is maintained.

only one-eighth appears above the surface. The small and the large iceberg will both project above the surface in the same proportion (Figure 13.5), and we can say that both are in a state of equilibrium. If we took some earth-moving equipment on to the larger iceberg and removed large quantities of ice to the smaller, we would not be in danger of drowning because as we removed part of the ice the rest would rise to maintain a one-eighth part above the water. Similarly the smaller iceberg would sink under the additional weight, and would not have more than one-eighth above the water.

The granitic crust is less dense than the basaltic crust and underlying Mantle, and may be said to be 'floating', like the icebergs in water, in a state of gravitational equilibrium that is known as **isostasy**. Where the continental rocks rise to greater heights they are compensated at depth by roots extending into the Mantle; if there is a shift in the load there will be movements to adjust the thickness of the crustal rocks, although, as they are taking place in solid rocks, they will take longer than those in the case of the iceberg floating in water. For instance, when a range of mountains is worn away a lot of debris will be produced which is taken to the sea by rivers and glaciers and deposited. The mountains have lost a large mass of material, which has been added to the sea-floor, and so we must expect the mountainous area to rise slightly to maintain the isostatic balance, and the sea-floor to subside under the extra weight, as shown in Figure 13.6. A similar process takes place when large masses of ice accumulate on the land, as they did in the Ice Ages. The weight of ice causes the land to sink, and when the ice melts the land rises again. Figure 13.7 illustrates what happened when Scandinavia was covered by a thick ice-sheet: the isostatic 'recoil' is still taking place and the floor of the Baltic Sea is rising to resume its position before the ice formed.

Although we have built up a general picture of what lies beneath our feet inside the Earth, and of one of the very important effects of this arrangement, it would be an advantage to be able to confirm our indirect observations by a more direct method. The Americans attempted an interesting project some years ago in trying to drill through the Crust to find out the actual composition of the Mantle. As

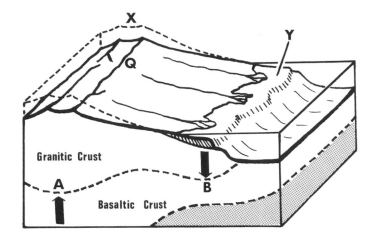

Figure 13.6 Isostasy (2) Erosion of mountains. Erosion and deposition lead to changes in the pattern of crustal thickness. The tops of the mountains at X are removed by erosion and the debris is deposited at Y. This leads to the removal of the load above A, which rises, and an increase of the load at B, which subsides. It is probable that some of the basaltic crust moves slowly from B to A.

Figure 13.7 Isostasy (3) Ice on Scandinavia. These generalised diagrams summarise the recent history of Scandinavia. 1 Shows the peninsula at the beginning of the Ice Age. The extra load of ice caused the crustal rocks to sag. 2 When the ice melted the sea rose and drowned the depressed area of land. 3 Today land in the Baltic Sea area is rising by a process of isostatic recoil: the load of ice has been removed, and the land is rising gradually to its former position.

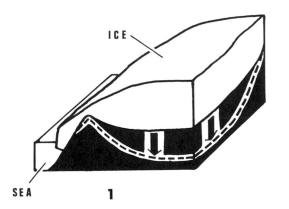

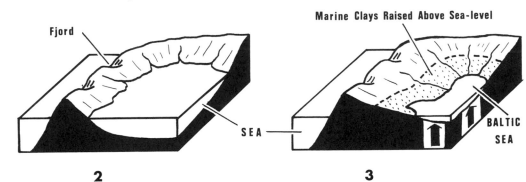

you have seen in Figure 13.4 the Crust is thinnest under the oceans and this is where they hoped to drill. The aim was to drill through the Mohorovičić Discontinuity and was known as the 'Mohole Project'. The engineering difficulties, which were immense, held it back, and the rising costs led to the cancellation of the programme in 1969. It is likely, however, that further attempts will be made in the future.

How Was the Earth Formed?

The answer to this question is bound up with the origin of the whole Solar System—of which the Earth is one member—and therefore with the science of Astronomy. We still understand so little about conditions in the vast expanses of space that our ideas about the origin of the Earth cannot be very definite. That is why there are so many theories about it.

For many years it has been assumed that the Earth cooled from a hot, molten mass, and that contraction took place as it solidified. Recent research, especially connected with radioactive minerals, has caused scientists to change their minds. One of the most widely accepted modern theories concerning the origin of the Earth and the Solar System generally suggests that the Sun was one of a series of stars developing out of a great contracting cloud of dust and gas. Some of the gases and dust left over formed a giant disc around the Sun, but this soon broke up again into large masses. These masses grew by the accumulation on them of many **planetismals** (formed of fine silicate dust, water and ammonia) to form our present **planets**: moons and meteorites may be the remains of planetismals which failed to join up with a planet. All this happened at relatively low temperatures, and the Earth was originally a cool body, solid throughout.

The next stage in the Earth's development is also one which has given rise to a lot of discussion based on the few facts available. Some scientists suggest that the original dust forming the Earth contracted as the small but important quantities of radioactive elements decayed, releasing energy and melting the central Core. These reactions also resulted in geochemical separations as the heavier minerals sank in the molten interior to form the heavy nickel-iron Core surrounded by

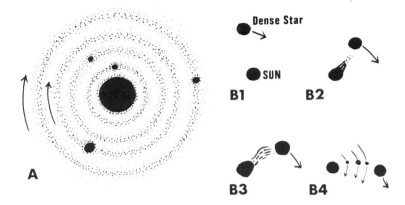

Figure 13.8 Two ideas concerning the origin of the Solar System. In A the planets are accreting out of a mass of dust and gas whirling around the proto-Sun. B shows a series of stages which might have occurred if a large star had passed close enough to the Sun to pull away some of the material into a filament; such a filament might then have broken up into planet-like masses some of which were still held in orbits around the Sun.

layers of silicate minerals of decreasing density. Water was then formed on the surface of the Earth by the condensation of water vapour. The remaining gases involved formed the first atmosphere composed of ammonia (NH_3), water vapour (H_2O) and methane (CH_4), but these broke up as sunlight decomposed them. Most of the hydrogen escaped into space, leaving a preponderance of nitrogen and oxygen. Volcanic action added a small proportion of carbon dioxide, which has been a most important ingredient for the development of plant life.

Many scientists feel that it was not until plant life became established on the Earth that oxygen entered the atmosphere. It is thought that in the earliest days all life was restricted to the sea, where it was protected from the ultraviolet rays of the Sun, which would have burnt it up. Oxygen was gradually liberated in increasing quantities by the processes of photosynthesis, until life became possible nearer to the ocean surface. The rate of oxygen accumulation increased markedly at this stage. Throughout the Precambrian the oxygen quantities available would supply only sufficient energy for the life processes (metabolism) carried out by single-celled creatures. By the beginning of the Cambrian period, or just before, a wide range of many-celled (metazoan) creatures existed, and it is thought that a critical level of oxygen content in the atmosphere was passed: 1 per cent of the present level is suggested.

The first land life emerged in the late Silurian period, nearly 200 million years later. At this stage there must have been sufficient oxygen to build up the protective ozone layer in the upper atmosphere, which absorbs the ultraviolet radiation as it enters the atmosphere. After this life could develop on the Earth to the stage where we find it today. This facet of the Earth's development illustrates the fact that life has not been evolving against a static background and in a static atmosphere, but that all these aspects of our planet's existence have been evolving together and affecting each other on the way.

How Old Is the Earth?

Over the last 200 years scientists have realised that the Earth is very old. We can all observe that if our local river is wearing down its valley it is doing it so slowly that we cannot see what is happening. Rivers become swollen and discoloured by mud after a storm, and this mud must have been washed into the stream by the rain. As it will never be returned to the field from which it was taken, the land surface has been lowered very slightly. Over many years the process becomes important, and it has been calculated that the surface of Great Britain is being worn down at a rate of 30 cm every 3 or 4000 years. Even this rate is rapid compared with the deserts, where the rocks are scarcely affected over periods of time involving tens of thousands of years. Egyptian monuments nearly 6000 years old are nearly as fresh as when they were made, whilst in Britain we have difficulty in maintaining buildings a few hundred years old.

This slow destruction of rocks and relief at the Earth's surface is part of the series of events which have affected the world since it was first formed. Periods of rock destruction which must have lasted for a very long time have been followed by the formation of new rocks and by the slow uplift of new mountains accompanied by the dramatic outpouring of volcanic lavas. As many as a dozen of these cycles have been recognised in some parts of the world.

Having realised that the earth is very old we would like to have some idea of its actual age in years. Scientists have estimated geological time in a number of ways. Some worked out the total thickness of rocks formed as sediments in the sea, estimated the time taken to form a certain thickness, and multiplied this figure by the total. If, for instance, we found a layer of mud 50 cm deep covering the wreck of a Roman galley containing dated coins, we might assume that 50 cm of mud takes, say, 2000 years to form. Our calculation of the total thickness of marine sediments in our local area might add up to 25 000 m, and by combining these figures we would reach a date for the formation of the first sedimentary rocks at 100 million years ago.

Another result was obtained by assuming that the oceans were originally fresh water which had condensed from clouds of gas, and working out how much salt is being supplied in river solution at the present time. Both of the answers were wrong because their authors did not make allowance for the facts that the rate of rock-formation and salt-accumulation in the oceans has varied throughout geological time, that a large quantity of salt has been precipitated out of the waters, and that both processes are going on at a faster rate than usual at present. These ages were thus too small.

The best methods of dating the rocks accurately involve using the results of a natural process which has operated at a known rate from a definite starting-point in time. Varves are formed in lakes from the debris brought in by glacial meltwater. The coarser material (silt and sand) settles quickly in summer and is followed by the finer clay in winter. Each varve represents a year's deposition. By counting such annual deposits in Sweden it has been calculated that ice retreated from southern Sweden 13 500 years ago, and the far north of Sweden

8700 years ago. This method, however, can only be used for short periods of time.

In recent years the science of nuclear physics has come to the aid of geologists. Some elements like uranium and thorium are unstable, and gradually break down into more stable elements, such as lead. There are very good grounds for believing that the rate of disintegration, which can be calculated, is not affected by external conditions such as heat, cold, pressure, chemical changes, etc, and has remained the same throughout geological time. If the amounts of uranium and lead are measured the age of the rock can be calculated. The uranium–lead method is a good one, but rocks containing the minerals are not common. On the other hand, the potassium–argon method uses a more common element, but is more difficult to handle.

The use of these radioactive mineral 'clocks' enables us to give reliable dates to rocks for the first time. The oldest rock occurring at the surface of the Earth is now known to be 3900 million years old, but the planet was first formed nearer to 4500 million years ago. The date of the planet's origin is unknown because the oldest rocks we can examine at the surface were certainly not the first to form: they are metamorphosed sediments and volcanic rocks, and the sediments at least must have been derived from even older rocks.

14 The Evolving Continents

The evolution of the Earth's atmosphere and the evolution of life on the Earth have been accompanied by, and are interdependent with, the evolution of the continental masses and their relief features. Our studies of the Earth's interior, the ocean floors and the rock-forming processes have shown that the continents are formed of material which is different from the oceanic crust. Whilst the ocean floor material originates largely in the upper Mantle, the rocks of the continents are formed by a wide range of processes, which are summarised in the Rock Cycle discussed at the end of Chapter 11. Reactions with the atmospheric gases—particularly oxygen—and alteration due to burial, as well as volcanic activity, have given rise to these distinctively continental rocks, which have a lower density and therefore stand out above the more dense ocean-floors.

This difference in composition accounts for many of the general differences between the ocean basins and continents, but not for the more detailed features, such as the shapes and present positions of the continents on the globe. A study of these latter characteristics has been a source of debate and speculation since the Earth was shown to be spherical and the continental positions became known. The answer is not only of interest to the curious speculator, but has important outcomes for a wide range of scientists who base their work on the processes of historical evolution: thus biologists seeking to explain the world-wide distribution of plants and animals, and geologists attempting to account for such phenomena as volcanic activity, earthquakes and mountain-building, are all bound to consider the early history of the major features of the Earth's surface.

There is no single, overwhelmingly conclusive piece of evidence concerning the evolution of the continents. Many fields of learning have contributed to this study over the years. Not only does the concept give us an overall view of the development of major Earth features, but the types of evidence are drawn from all facets of the science of geology and beyond.

The Geographical and Geological Evidence

Almost as soon as the first maps of the world were produced by Mercator (1569) people began to speculate on the relationships and patterns made by the shapes of the continents. Lilienthal, a German theologian, noted in 1756 that the opposite sides of the Atlantic Ocean fit together very well: this is particularly clear in the South Atlantic, where the east coast of South America and the west coast of Africa can be placed together with very few gaps or overlaps. In the twentieth century geologists have returned to this, but have decided that the present coastline is not such a good guide as a line round the continental slope: Figure 14.1 shows this fit. In fact the fit is more than

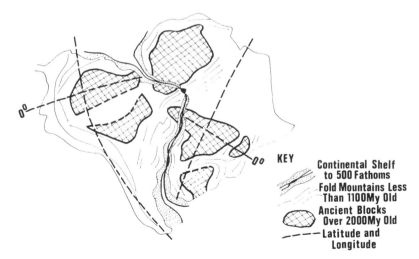

KEY
- Continental Shelf to 500 Fathoms
- Fold Mountains Less Than 1100My Old
- Ancient Blocks Over 2000My Old
- Latitude and Longitude

Figure 14.1 The fit between South America and Africa. How closely do the coastline and the geological features match up across the Atlantic Ocean?

merely geographical: small fragments of extremely ancient (over 2000 million years old) continental basement rocks (known as cratons) have been isolated on the north-east coast of South America, but fit in as extensions of larger outcrops of these ancient rocks in Africa. The trends of more recent mountain ranges also continue across the join. In addition both continents had very similar geological histories until the end of the Cretaceous: marine deposits of late Jurassic age occur along the margins between them, indicating that the break was beginning, but no deposits older than the Cretaceous are known from the Atlantic floor.

A number of difficulties concerning the distribution of climatic zones in the past have long puzzled geologists. The southern continents (Africa, South America, Australia, Antarctica and peninsular India) all show evidence of a glaciation in the late Carboniferous and Permian times (Figure 14.2). This group of continents is known as **Gondwanaland** because they have so many common geological simi-

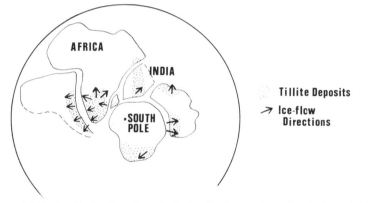

Figure 14.2 The Carbo-Permian glaciation in the southern hemisphere. This map shows the evidence for this event in terms of tillite deposits (ancient boulder clays) and striated rocks beneath these. It seems necessary to bring the southern continents together into an area which experienced the coverage by ice sheets.

larities. Whilst these continents were experiencing a glaciation, rocks of the same age in Britain, and as far north as Spitzbergen, are thick coal seams, which can only have been formed in wet, tropical swamps, and these are followed by thick layers of salt formed by intensive evaporation in hot desert conditions. Coral reef limestones are also found in these northerly regions—features which are now forming only in tropical waters.

In addition many of the features of plant and animal migration in the past can most easily be accounted for if continents, now separated by wide, deep oceans, were once joined. The Gondwanaland continents, for instance, have a unique group of reptile and plant fossils.

So much was known before 1960, although more recent studies have confirmed the evidence and added dates based on radioactive minerals. Many scientists, however, found it difficult to imagine that the hard rocks of the Earth's Crust could have moved over the hard rocks of the Mantle. This would have been necessary if the continents drifted to their present positions from one where they were joined to each other. Lines of evidence began to mount to confirm the idea that large-scale horizontal movements can take place: we saw in Chapter 11 that the seaward side of the San Andreas fault in California has moved nearly 500 km to the north in the last 180 million years.

The Geophysical Evidence

Geophysicists are scientists who study the physical properties of Earth materials. They have been particularly associated with the study of the Earth's interior, but have also made investigations into the Earth's magnetic and gravitational fields and into the significance of the radioactive minerals in the Earth's crustal rocks. Although it was the geophysicists who raised objections to the geological evidence for continental drift before the 1950s, they have led the way since then in establishing the concept as one of the most important general theories in modern Earth Science.

Work on **palaeomagnetism**, using the rocks as 'fossil compasses', began only in 1950 and, although the techniques used are not yet perfect, much has been discovered which is of great interest. The

Earth's magnetic field has a very similar effect to that of a great bar magnet near the Earth's centre and orientated north–south along the Earth's axis. Our studies of the Earth's interior showed that the structure is a series of concentric shells, and so there is no actual bar magnet as such: the effect must be caused by another means, but we have little evidence concerning this at the moment. Our observations also tell us that the magnetic poles are not the same as the geographical North and South Poles—the difference between the bearings of the magnetic and geographical poles being known as the angle of magnetic declination—but over a period of several thousand years the average position of the magnetic poles is thought to be very close to the geographical axis.

As magma cools tiny crystals of magnetite (iron oxide) begin to form. After they have become solid, the cooling process reaches a stage at which these minerals acquire a direction of magnetisation controlled by the Earth's magnetic field at that time—ie they present us with a 'frozen' record of the Earth's magnetic field in past times. Figure 14.3 shows the effect you get if you do an experiment with a series of small compasses and a bar magnet. The angle of magnetic dip, or inclination, is the angle between the axis of the needle and the horizontal (defined as a tangent to the sphere at that point). It can be found approximately by using a 'dip needle' which your school physics department may possess. Check the result by a simple calculation: the tangent of the angle of magnetic inclination is equal to twice the tangent of the latitude at which the observation is made. Thus at 30 degrees North the angle of magnetic inclination is 49 degrees. Figure 14.4 shows the results of measurements taken across a Hawaiian lava field. A visit to the new volcano, Surtsey, off the coast of Iceland, showed that the newly formed lavas contained iron-rich minerals orientated in accordance with the present magnetic field of the Earth.

Many measurements of rock magnetism have now been made all over the world. Of course the measurers have to be very careful to notice the exact position of the rock and its orientation as they collect it: if possible they like to collect specimens from a whole range of lava flows erupted over a period of thousands of years, so that they can

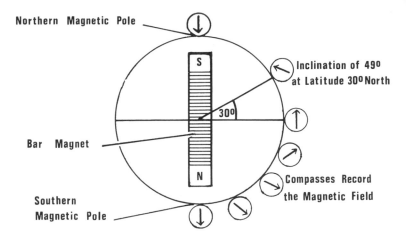

Figure 14.3 A magnetic field. This shows the result of an experiment you can try yourself, which illustrates the effect of the earth's magnetic field on iron minerals in the rocks. Place a petrie dish over the magnet. The attitude of a mineral is determined by the part of the Earth where the rock which contains it solidified.

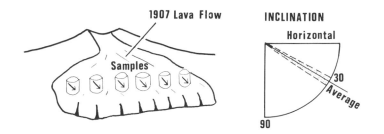

Figure 14.4 An illustration of the methods used in palaeomagnetism. Cores of lava were drilled from a flow which took place in Hawaii in 1907. The direction of magnetic inclination calculated from these agrees with the magnetic field for that date.

assume that the average positions of the magnetic poles were at the geographical poles. The results have shown that, although the magnetic poles have been very close to the geographical poles over the last 20 million years, they tended to wander before that: a Permian pole was in the centre of the northern Pacific Ocean. When polar wandering curves are compared for different continents (Figure 14.5) it is seen that each continent has its own. This means that the poles and continents have moved relative to each other, the extent of the movement differing from continent to continent. It seems inescapable, therefore, to conclude that movements of the continents themselves account for at least part (and probably most) of this relative movement.

The end of the story is not in sight, but enough is known to confirm the idea that the Gondwanaland continents were grouped together round the geographical South Pole in the Carboniferous and Permian periods, and that Britain was near the Equator at the time when extensive coal deposits were being formed.

The features of **ocean floor relief**, described in Chapter 6, have also been studied in relation to their magnetic features: rapid surveys have been made possible since 1960 by towing magnetometers behind survey ships, the patterns of alternate high and low values being mapped in bands parallel to the oceanic ridges. The examination of lavas erupted on land areas above these ridges (eg Iceland, California) has shown that there has been a series of reversals of the Earth's magnetism over the last $3\frac{1}{2}$ million years: approximately every million years the north magnetic pole becomes a south magnetic pole and vice versa. These reversals seem to be associated with the alternating magnetic stripes on the ocean floor and with the oceanic ridges. The fossil magnetism of deep ocean sediments has confirmed this association, since they show a similar sequence of events. Figure 14.6 summarises the results of these studies, which suggest that new ocean floor is created at the oceanic ridges and spreads outwards. The rate of spreading can be worked out by estimating the date of rock formation and its distance from the centre of the ridge: such rates vary from averages of 1 cm/year near Iceland to as much as 9 cm/year in the east Pacific. The interesting result of this work is the conclusion that about half of

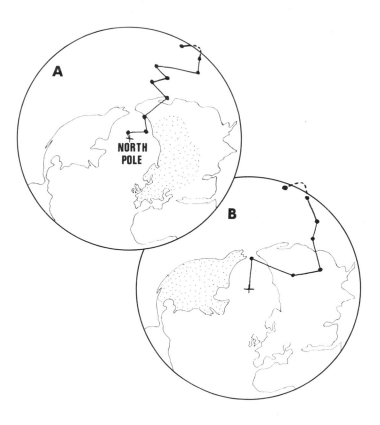

Figure 14.5 Polar wandering. Measurements of palaeomagnetism in ancient rocks from a selection of sites across Europe and northern Asia suggest that the position of the magnetic pole has moved as shown (A). The pattern established in a similar way for North America (B) is somewhat different: each continent has its own path of polar wandering. Of course it may have been that the continents have wandered with respect to the poles, rather than the other way round!

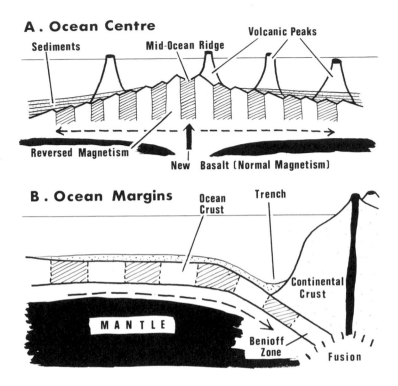

A. Ocean Centre

Sediments   Mid-Ocean Ridge   Volcanic Peaks

Reversed Magnetism

New Basalt (Normal Magnetism)

B. Ocean Margins

Ocean Crust   Trench

Continental Crust

MANTLE

Benioff Zone   Fusion

Figure 14.6 Two aspects of spreading sea-floors. In the central parts of the oceans (or at the oceanic ridge) new lava is injected from beneath; both the igneous rocks and ocean floor sediments increase in age from the ridges towards the margins of the oceans (A). The ocean margins (B) are often zones of volcanic and earthquake activity—particularly marked in the Pacific where the spreading movements of the floor are most rapid. This diagram shows one interpretation of what is happening.

the ocean floor has been formed in the last 65 million years, and no part, even of the largest ocean, the Pacific, is known to be older than the Jurassic period (170 million years). Thus the ocean floor rocks are much younger than the continental rocks.

Our evidence so far suggests that both the continental masses and the ocean floor rocks are subject to horizontal movements over the Earth's surface. Figure 14.7 summarises the changes in the positions and shapes of the continents and major islands which are thought to have taken place since the end of the Palaeozoic. The evidence for these movements, and the youth of the ocean floors has largely obscured the evidence for earlier movements, but it is thought that these occurred.

The greatest problem remaining is to find a mechanism which could have given rise to these changes. One idea suggests that the Earth's crust is composed of large, rigid plates which are being formed at the ocean ridges and consumed as they plunge beneath the ocean trenches or continents. The volcanic and earthquake activity associated with these trenches is explained as due to the stresses involved in the digestion by the Mantle of the descending oceanic crust. Figure 14.8 shows the major plate areas. It has also been suggested that a zone in the upper Mantle between 100 and 200 km deep, known as the Low Velocity Zone (or Asthenosphere) because it transmits the earthquake shock waves more slowly, is a zone of relative weakness where high temperatures may locally overcome the effect of increasing pressure with depth. If this zone can act as a relatively viscous (stiff liquid) zone between the rigid Crust above and the rigid Mantle below, it might account for the observed surface movements of the continents.

Many questions remain unanswered concerning the motion of these plates. How do they move? What is their relationship to the internal structure of the Earth? Has the Earth always been the same size? Is it expanding or contracting? All these questions are important in the scheme expounded here, but the answers are not at all clear at the moment. And yet the concepts of continental drift, ocean floor spreading and plate tectonics have given rise to explanation patterns concerning a variety of Earth phenomena: earthquake and volcanic activity,

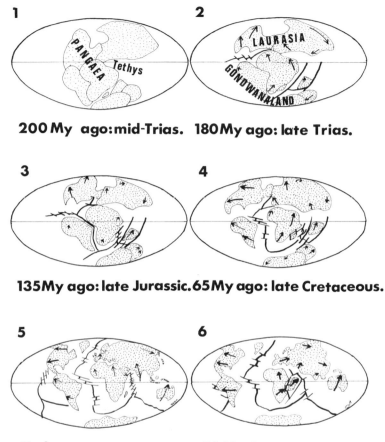

1 **2**

200 My ago: mid-Trias. 180 My ago: late Trias.

3 **4**

135 My ago: late Jurassic. 65 My ago: late Cretaceous.

5 **6**

Today **50 My hence?**

Figure 14.7 The generalised development of the continents of the Earth's surface during the last 200 million years. These maps are based on a wide range of observations, but also include some reasoned guesswork.

the generation of the main types of crustal rocks and the formation of mountain ranges.

The Origin of Mountains

In Chapter 12 we discussed some of the major features of the ranges of folded mountains which form the main relief features of the continents. Figure 6.5 shows that the present high mountain ranges are located in two zones:

(1) The Alps–Himalayas chains lie between two continental masses: Eurasia to the north and Africa–Arabia–peninsular India to the south.

(2) The circum-Pacific ring which includes the Andes, the western cordilleras of North America and the islands and island arcs of the west Pacific. These are along the margin between continent and ocean.

Both groups are along the margins of plates as shown in Figure 14.8. The series of events in which sediments of great thickness are deposited, folded, thrust and uplifted is due to the convergence of plates. This leads first to downwarping of the Crust and the accumulation of sediments, but as compression increases they are folded and uplifted. The unusual flow of heat to the surface at these zones of crustal plate convergence leads to regional metamorphism and igneous activity (Figure 14.9). Once two continental plates come together in this way they become welded, Earth activity dies down and the zone is changed into a stable part of the Crust.

The present chains are the most recent of a series which have been raised since the beginning of geological time: Figure 12.18 gives the dates of those recognised in Europe. Since the Cambrian the phases of mountain building have lasted approximately 200 million years, and it has been suggested that this is a fundamental process affecting the Earth's Crust. The most recent cycle, leading up to the present dispositions of the continents and the raising of the Alpine mountain ranges, began at the end of the Palaeozoic. In order to search back beyond this time we need to study the older mountain ranges, which suggest that other continents may have drifted together to cause these orogenic zones. The map of the North Atlantic area (Figure 14.10) shows some of the evidence involved in this sort of exercise.

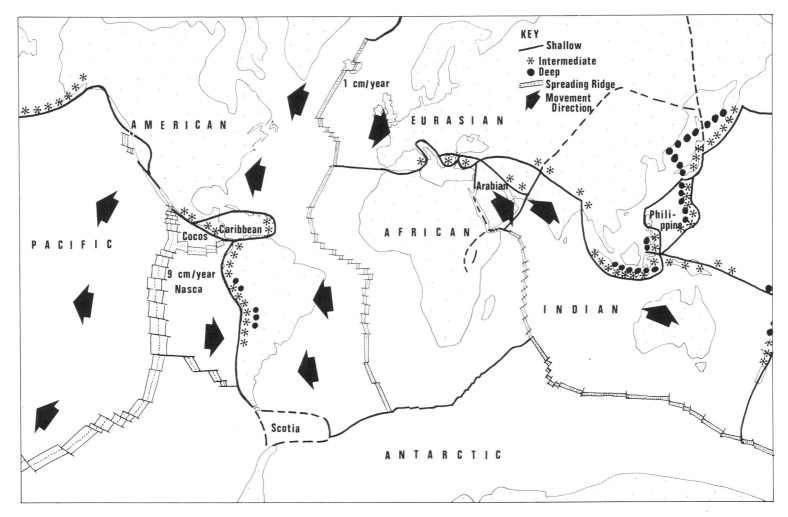

Figure 14.8 Major world plates related to earthquake depths.

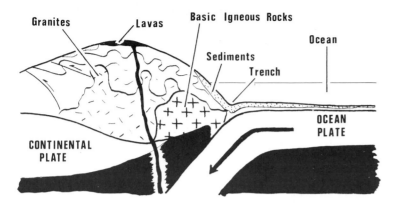

Figure 14.9 Mountains and plates. Folded mountains formed where an oceanic plate dives beneath a continental mass.

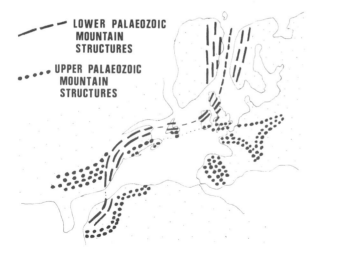

Figure 14.10 North Atlantic 'fit'. Older mountain chains related to the closed North Atlantic. Perhaps these chains were the sites of former oceans.

It seems most likely that the Earth's surface has witnessed the movements of continental masses across its surface in the past. At times they combined into a great whole, but at others they were parted into separate masses. We thus build up a picture of the continents combining and recombining down through the history of the evolution of the Earth's surface—just as we formed the picture of the rock materials combining and recombining through the progress of the Rock Cycle.

15 Earth History: Principles

When we examine a quarry or cliff section we are looking at the results of geological processes which began many millions of years ago and which have affected these rocks at intervals since. The observations we make may show that a group of sedimentary rocks were deposited over a certain length of geological time and under particular conditions; at a later date the soft sediments were hardened and fossil shells dissolved away to leave moulds and casts; and there may be evidence of folding or tilting with uplift. This type of evidence, tied in or correlated with other evidence from farther afield, has given rise to a detailed knowledge of the past evolution of the British Isles and other areas of the world. The study of the layered arrangement of the rocks, or strata, is known as **stratigraphy**.

Many of the principles we apply to this study of the past seem simple and even obvious; many are similar to principles and methods of working in other sciences and even historical studies. And yet there was a long period of struggle, which lasted into this century, before they became accepted. Our ancestors, and particularly those who lived before 1850, would not have accepted these ideas so willingly.

Early Ideas

Man's earliest writings show that he has always been interested in his past and origins, and at various times in his history he has related the evidence he has found in the rocks to these problems. Greek philosophers as early as 500 BC had decided that what we now call fossils had once been living animals, but like many of their discoveries the practical implications were not clear at the time. True geological studies were held up until the mid-seventeenth century AD by three major factors.
(1) Observers used **the wrong method**. They made a few, often uncon-

The Moon surface. Describe the features shown in this photograph (the large crater in the foreground is 10 km across). How do these differ from the Earth? Can you explain these differences? Consider the fact that the rocks so far examined at the surface of the Moon are older than nearly all the rocks at the Earth's surface.

nected, observations from nature and spent many hours speculating on their possible origin, often incorporating concepts which could not possibly have been observed or measured. Thus if something was beyond explanation at a particular moment they would bring in the supernatural. This approach brought with it a series of strange ideas and misconceptions, as the following passage, written in the mid-fourteenth century, shows:

It often happens that in one place or another the earth shakes so violently that cities are thrown down and even that one mountain is hurled against another mountain. The common people do not understand why this happens and so a lot of old women who claim to be wise say that the earth rests on a great fish called Celebrant which grasps its tail in its mouth. When this fish moves or turns the earth trembles. This is a ridiculous fable and of course is not true. . . . We shall therefore explain what earthquakes really are and what remarkable consequences result from them. Earthquakes arise from the fact that in subterranean caverns and especially those within hollow mountains, earthy vapours collect and these sometimes gather in such enormous volumes that the caverns are not able to hold them. They batter the walls of the cavern in which they are and force their way out into another and still another cavern until they fill every open space in the mountain. . . . If they cannot reach the surface they give rise to great earthquakes. This unrest is brought about by the mighty power of the stars. (*Conrad von Megenberg*, d. 1347 AD)

(2) Geological studies are based on a **correct knowledge** of chemical, physical and biological processes, and it was not until these sciences began to understand the true nature of their materials that advances could be made in geology. Rock-forming minerals were only analysed chemically in the nineteenth century; the discovery of radioactivity did not take place until the very end of the nineteenth century; and the theory of evolution by natural selection was a mid-nineteenth century proposal. All have made important contributions to our understanding of the past. (3) There was a tendency to confuse a science dealing in origins with **religious beliefs**, and even with unwarranted additions which have been made to these beliefs. The belief that the Earth's history was confined to 6000 years was not a matter of Biblical interpretation and simple mathematics, but was the legacy of an early Christian administrator who liked rounded numbers! And yet it handicapped and restrained the progress of geology for hundreds of years. It was only when it could be seen that there was a science of geology based on observation and measurement, subject to the limitations of the observers and their instruments, and distinct from the search for meaning and purpose in life (the province of religion), that progress was possible.

The Beginnings of Modern Geology

Nicolaus Steno (1638–87), a Danish priest, put forward the basic tenets of stratigraphical studies in 1669. In conflict with the 'Flood geologists' who held that the rocks were formed when a mass of sediment settled out of the declining waters of Noah's flood, he stated that each layer of rock was formed on top of earlier strata by a separate series of sedimentary processes: the lowest layers are therefore the oldest. This idea seems obvious to us, but he had to give it an imposing title: *The Law of the Superposition of Strata*. This principle is applicable in each local area, although there are areas where intense folding causes some of the rocks to be upside down. We may be able to use internal structures, such as cross-bedding or graded-bedding to determine whether the rocks are the right way up or not. On the whole the oldest rocks have the fewest fossils and the greatest degree of folding. We have been into these factors in detail in Chapters 8 and 12. Other principles which Steno referred to in his work include:

1 **The Principle of Original Horizontality.** Most sediments and lava flows were originally horizontal (Chapter 12), although there are exceptions to this rule which should be kept in mind (Chapter 8).

2 **The Principle of Original Lateral Extension.** This states that rock layers formed at the same period in time can often be traced over a wide area, although these layers may be cut into and separated by erosion.

3 **The Principle of Cross-cutting Relationships.** Igneous intrusions (Chapter 10) and unconformities (Chapter 8) give rise to situations where one group of rocks cuts across another. The rocks which cut across another group are always younger in age.

Steno thus provided a basis for the study of rock sequences, although his conclusions were only being applied for the first time 100 years later. This study of rock sequences was useful in local areas, but widespread correlation on the basis of rock-type alone is not satisfactory, since different types of sediment were laid down at the same time. William Smith, an English engineer, journeyed about the country in the course of examining the ground as a prelude to canal- and bridge-building. He received no formal geological education and was thus free from ideas such as Flood Geology, which were still being taught at the end of the eighteenth century. He found that, although a rock of a certain age might change its nature from sandstone to clay over a distance, it would still contain some of the same fossils (Figure 15.1). Using this principle of **dating the rocks by the fossils they contained**, he was able to publish a very reasonable geological map of England and Wales in 1815, bearing many resemblances to the maps of today. Many of his names for the rock horizons are still in use. He anticipated Darwin's ideas on evolution, which were the basis of his own practical scheme of working, by sixty years.

Whilst the sequence of rock layers could be established and correlated with other sequences in different parts of the country, there was also the matter of interpreting the conditions which gave rise to them: what were the geographical situations at a particular period in the past? Flood Geology suggested that all the present dispositions of the rocks had been formed at one moment in time. Another interpretation, associated with the famous German teacher of the late eighteenth century, Abraham Gottlob Werner (1749–1817), argued that all the rocks had been formed in the sea: granites were chemical precipitates, as was shown by their crystalline nature, and rocks like those in the Giant's Causeway were enormous crystals formed by chemical precipitation in the sea. Werner's ideas became associated with Flood Geology, and the religious note kept them alive until the mid-nineteenth century, but neither were based on close observation of the evidence in the rocks. Werner never travelled out of Upper Saxony, but as soon as his own pupils began travelling farther afield, visiting ancient volcanic centres such as the Auvergne of central France and the Inner Hebrides of north-west Britain, they became convinced he was wrong.

James Hutton (1726–97) started the modern approach to the interpretation of past events as shown by the rock evidence. Not only did he conclude that some rocks had cooled from a molten state, and so took up sides against Werner in the debate of that time, but he went further. Even more important for the development of geology was his concept that the forces we observe acting on the landscape today have been in operation throughout geological time; we do not need to suggest that any other forces have operated in the past, which we cannot observe today. He expressed this in the following way:

In examining things present we have data from which to reason with regard to what has been; and, from what has actually been, we have data for concluding with regard to that which is to happen here after. Therefore, upon the supposition that the operations of nature are equable and steady, we find, in natural appearances, a means of concluding a certain portion of time to have necessarily elapsed, in the production of these events of which we see the effects. . . .

But how shall we describe a process which nobody has seen performed, and of which no written history gives any account? This is only to be investigated, first, in examining the nature of those solid bodies, the history of which we want to know; and secondly, in examining the natural operations of the globe, in order to see if there now exist such operations as, from the nature of the solid bodies, appear to have been necessary in their formation. . . . Therefore, there is no occasion for having recourse to any unnatural supposition of evil, to any destructive accident in nature, or to the agency of any preternatural cause, in explaining that which actually appears.

His was a truly scientific approach, limited to the observations he could make, and he could find no evidence to suggest that geological processes had not operated from the earliest times:

But if the succession of worlds is established in the system of nature, it is in vain to look for anything higher in the origin of the earth.

The result therefore, of our present enquiry is, that we find no vestige of a beginning,—no prospect of an end.

Hutton's friend, John Playfair, popularised his views in the early years of the nineteenth century. In 1830 Charles Lyell, who had recently been persuaded that Hutton was right, published the First

Edition of his book, *Principles of Geology*, which was to be the basic text for geologists during the next fifty years and greatly influenced Charles Darwin's view of the world. Its subtitle: 'Being an attempt to explain the former changes of the Earth's surface by reference to causes now in operation', shows how it followed Hutton's basic theme. It has been said that Lyell's *Principles*, and Darwin's *Origin of Species* created a revolution in scientific thought in the nineteenth century akin to that associated with Newton, Kepler and Galileo in the seventeenth.

The basic concept for interpreting the rock record put forward by Hutton and elaborated by Lyell earned the nickname '*uniformitarianism*' from its opponents, and was later summarised in the dictum: '*The present is the key to the past.*' The concept has maintained its place as the basis of geological studies despite continuing criticisms of its viability. Physical, chemical and biological processes have acted in the same ways over the hundreds of years covered by written historical records, and this lends support in the short term. Over periods corresponding to the geological time-scale one can point to the fact that light from a nebula, which we believe started its journey to the Earth hundreds of millions of years ago, contains the same characteristic group of lines as is seen today in a hydrogen spectrum in the laboratory. This suggests that the processes may have acted in the same ways throughout time. The differences in the geography of the past—ie fluctuating ice-sheets, moving continents, changing positions of mountains, completely different groups of animals and plants—do not affect the case, since they can all be seen as part of the grand scheme of a planet developing under the processes which are still in operation.

James Hutton and William Smith can thus be regarded as the true fathers of geology and stratigraphy in their modern aspects. Before 1780 there had been long ages of groping in the dark for the keys which would unlock the door to the ideas needed to interpret the mass of geological facts gathered over hundreds of years. By 1830 Charles Lyell could produce a systematic account of the geological information which was to hand in terms which are relevant today. The way was open for the research which has led to our modern understanding of the geological history of our lands. Playfair wrote the following in his Introduction to the *Illustrations of the Huttonian Theory of the Earth* (1802) which began the revolution:

. . . we shall see abundant reason to conclude that the earth has been the theatre of many great revolutions, and that nothing on its surface has been exempted from their effects.

To trace the series of revolutions, to explain their causes, and thus to connect together all the indications of change that are found in the mineral kingdom, is the proper object of a THEORY OF THE EARTH.

But, though the attention of men may be turned to the theory of the earth by a very superficial acquaintance with the phenomena of geology, the formation of such a theory requires an accurate and extensive examination of these phenomena, and is inconsistent with any but a very advanced state of the physical sciences. There is, perhaps, in these sciences, no research more arduous than this; none certainly where the subject is so complex; where the appearances are so extremely diversified, or so widely scattered, and where the causes that have operated are so remote from the sphere of ordinary observation. Hence the attempts to form a theory of the earth are of very modern origin, and as, from the simplicity of its subject, astronomy is the oldest, so, on account of the complexness of its subject, geology is the youngest of the sciences.

The last statement may raise a smile, but it illustrates the feeling that these early geologists had that they were at the beginning of great discoveries. Geology is still a growing, young science, and advances in our knowledge have been tremendous over the past twenty years. It has been calculated that 90 per cent of the geologists who have contributed to our knowledge are still living.

The Geological Column

In Chapter 2 you were introduced to the table of periods of geological time, which is often known as the Geological Column (Figure 2.6). It shows the relationships of the geological time units which were built up in the early years of the nineteenth century. At first they were based on the major rock-types and the obvious unconformity divisions between. Thus the Old Red Sandstone of the Welsh borders could be distinguished from the overlying grey limestones and coal seams included in the Carboniferous division, and the brick-red New Red Sandstones on top of the Carboniferous. The rocks which covered the New Red Sandstone were fossiliferous clays, sandstones and limestones, and these were named after fine exposures found in the Jura Mountains of France—the Jurassic system. Above these was another similar group including the distinctive white chalk, and this system of rocks was named the Cretaceous (Latin *creta* = chalk).

The youngest rocks of all were named after the uppermost of an earlier division of all rocks into Primary, Secondary and Tertiary groups. In 1833 Charles Lyell put forward a threefold division of these Tertiary rocks based on the number of fossil groups having living representatives. Thus the Eocene ('dawn of recent time') had 1–5 per cent of present-day species of animals; the Miocene ('moderately recent time') had 20–40 per cent; and the Pliocene ('most recent time') 50–100 per cent. Later the Pleistocene ('extremely recent time') replaced part of the Pliocene system, and the Oligocene ('slightly recent time') had to be introduced between the Eocene and Miocene.

More widespread studies of the two Red Sandstone formations produced one or two further modifications. The New Red Sandstone rocks were almost uniform in Britain, but on the Continent they could be divided into two major systems. Murchison and Sedgwick, two famous British geologists, were invited to Russia by the Tsar and named a thick group of fossiliferous limestones the Permian after the province in which they had seen them. In Germany the uppermost New Red Sandstone rocks could be divided according to three distinct rock types, and the whole group was named the Triassic. Before journeying to Russia Murchison and Sedgwick had visited Devon and Cornwall and found rocks older than the coal-bearing strata, but bearing little relationship to the Old Red Sandstone. Studies in the Rhine Slate Plateau of Germany confirmed that these Devonian rocks were the marine equivalents of the Old Red Sandstone land deposits.

The rocks beneath the Old Red Sandstone had posed greater difficulties, since they were very similar in lithology, were often highly folded and faulted, and contained few fossils. Murchison went to South Wales and worked down from the Old Red Sandstone base. He made two divisions, the Lower and Upper Silurian, named after the Silures tribe which once inhabited the area. Sedgwick worked in North Wales without any upward level for reference, and called the ancient rocks in that area the Cambrian system (Cambria = Wales). The discovery that Sedgwick's Upper Cambrian and Murchison's Lower Silurian contained identical fossils led to a bitter quarrel between these two friends, and it was not until after their deaths that Lapworth, who had worked out the complex geology of the Scottish Southern Uplands by using the graptolite fossils, proposed the name Ordovician (after a north Welsh tribe) for the overlapping rocks.

The Cambrian rocks, however, were the oldest in which any fossils could be found at the time, and all the most ancient schists and gneisses beneath them came to be known collectively as the Precambrian.

Thus the major divisions of geological time were established and given names with a variety of origins. It is amazing that they have stood the test of time and world-wide extension, and have been adapted to local conditions everywhere. The main modification has been in North America, where the Carboniferous is divided into the Mississippian (limestone) and Pennsylvanian (coal-bearing) systems. It has been found, however, that the original lines of division, often based on major British unconformities and rock-types, have not held true, and that fossil boundaries are more reliable. The general classification proposed by Philips in 1840 has been adopted:

The Cainozoic ('recent life')—Eocene to Recent.
The Mesozoic ('middle life')—Triassic to Cretaceous.
The Palaeozoic ('ancient life')—Cambrian to Permian. These three eras are now known together as the Phanerozoic ('evident life').
The Precambrian rocks contain few fossils, and may be divided into two groups—the most ancient Azoic ('no life') and the Cryptozoic ('hidden life').

These are the major **eras** of geological time, to which **periods** like the Cambrian and Cretaceous belong. The rocks formed during these periods are known as the members of, for instance, the Cretaceous **system**.

Fossils are the most important means whereby one rock is shown to be older or younger than another—ie given its age relative to other rocks—but they cannot give us the absolute age of rocks in years. This can now be done by using **radioactive minerals** contained in the rocks. The basic principle of this method was discussed in Chapter 13. The most reliable dates are those from igneous rocks, since the date recorded by such minerals is that at which the mineral was first formed (Figure 15.2). We can now prove that although the granite composing Lundy Island in the Bristol Channel is so close to Dartmoor (280 million years old), it is of the same age as much more recent granites in Northern Ireland and western Scotland (50 million years old). Sediments on the other hand contain minerals from a variety of rocks, and these may record several 'radioactive dates' covering a wide period of time. Certain minerals, like glauconite, are formed as the sediment is accumulating in the sea, and also have tiny quantities of radioactive elements which allow a sediment to be dated more accurately. Metamorphic rocks also present problems, because new minerals are often formed in the processes of alteration, but many metamorphosed rocks record several periods of change and thus several 'radioactive dates'. These are often significant in recording several phases of alteration and earth movement in an area: individual minerals within the rock give later dates than the rock as a whole, and this suggests that a later phase of metamorphism affected some minerals, but not all of the rock. This is very important in studying the Precambrian rocks.

A Summary of the Principles of Stratigraphy

(1) Our basic rule is **The Law of the Superposition of Strata**. We must first determine the order in which the rocks were formed. **Unconformities** are particularly important features of the rock succession, since they record major breaks in deposition due to uplift and erosion of the rocks formed previously. This gives us our local sequence of rocks in order of formation.

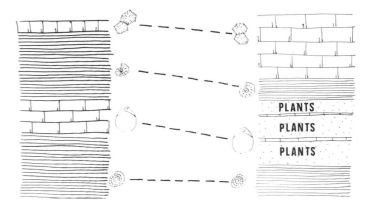

Figure 15.1 Correlation by fossils. Two successions of rocks 400 km apart have different thicknesses of rocks, but contain similar, distinctive fossils at several levels to enable a correlation to be made.

(2) **We correlate various local sequences mainly by the fossils** they contain if they are of Cambrian age or younger. This gives the rocks a series of relative ages (ie one rock is older, or younger, than another). (3) If the rocks do not contain fossils we can work out the sequence by various criteria showing which is the right way up. **Radiometric dates** based on radioactive minerals give us numerical dates for rocks and are especially useful where the rocks are of Precambrian age, since there are not enough fossils to enable correlation to be carried out. (4) When we come to **interpret the conditions** in which rocks were formed we use our knowledge of the present geological processes and compare their products with the rocks. We can tell whether a sea was advancing or retreating; whether the rocks were involved in a full orogenic cycle; or sometimes we can work out the prevalent climatic conditions on the continents of the times.

Stratigraphy and Economic Geology

Just as almost every other aspect of our geological studies has had an

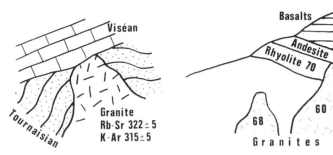

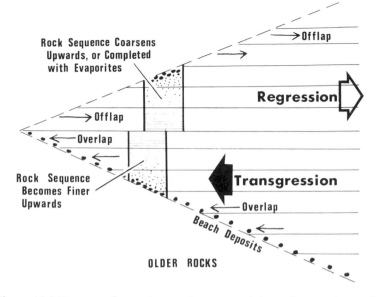

Figure 15.2 Radiometric dating. Radiometric dating is particularly successful with igneous rocks crystallised freshly from molten rock. The Vosges granite (A) has been used to give a date for the division between the lowest Lower Carboniferous rocks (Tournaisian) and those lying on top (Viséan): the granite intrusion altered the Tournaisian rocks, but the Viséan were deposited later, after erosion exposed the granite. Two radiometric dates were possible with this rock: the rubidium–strontium and potassium–argon methods. The Tucson Mountains (B) form a good example of the use of radiometric methods to determine the ages of a sequence of lavas and other igneous rocks without fossils.

Figure 15.3 Transgressing and regressing seas and the rocks associated with each process. Such changes in the distribution of sea and land are commonly recorded in the geological succession.

application in terms of an economic use, so the methods of stratigraphy are finding increasing uses in the search for new sources of minerals. Rocks formed in particular environments in the past often form distinctive oil reservoirs. Thus many of the Libyan oilfields occur in Cainozoic shoreline sands, and the Canadian oilfields of Alberta in Devonian reefs. The geologist in these areas looks for associations of rocks which suggest that such conditions occurred in the past.

Correlation of rocks by fossils is another method used for tracing likely oil-bearing formations across country. In addition there are some resources which occur in formations of a particular age. Most of

our oil comes from rocks formed in the last 65 million years, whereas most of our coal is obtained from rocks formed between 320 and 250 million years ago; Britain's chief brick clay horizon is the Oxford Clay of Upper Jurassic age, and most of the cement is made from the Upper Cretaceous Chalk.

An Introduction to the Historical Evolution of the British Isles

Geological time is divided into a number of periods, and each system of rocks formed during these periods has distinctive characteristics. We are fortunate that there are rocks of almost every period of geological time within the relatively small area of the British Isles. Only the Miocene system seems to be absent. These periods are largely bounded by time-planes determined by the fossils they contain, and so the fossil content of the rocks of different ages is a vital distinction between them. At the same time it is possible to trace a series of major events which affected this small section of the Earth's surface in terms of cycles of rock formation and mountain-building. There have been three of these in the last 600 million years since the beginning of the Cambrian. It will be a great help to keep this main outline of events before us.

The first of such cycles occupied the Lower Palaeozoic era (Cambrian, Ordovician and Silurian periods), when a deep geosynclinal trough extended across the centre of Britain and north-eastwards to the Scandinavian Peninsula. For nearly 200 million years sedimentation proceeded, accompanied in the Ordovician by volcanic activity and locally interrupted by minor earth movements. At the end of the Silurian intense folding and faulting were followed by the erection of high ranges of mountains stretching across the country at the beginning of the Devonian period. The mountains covering central and northern Britain were worn down by erosion during this period.

The second followed, but the geosyncline on this occasion was across the extreme south of the country and was connected to northern France and central Europe. Rocks were formed in this trough from the Devonian period onwards. In the succeeding Carboniferous shallow seas spread northwards to cover much of England and extended periodically into the Midland Valley of Scotland, being replaced later by deltas and coal swamps. This phase lasted about 200 million years and once again worked towards a climax of mountain-building at the end. On this occasion, however, the east–west ranges of fold mountains were confined to south Wales and England south of the Bristol Channel–Thames line, and the rest of the country was only mildly affected by the movements. At the end there was once again the erosion of mountains and the deposition of continental sediments (Permian and Triassic systems).

The third cycle was characterised by the inundation of much of England by shallow shelf seas during the Jurassic, Cretaceous and early Tertiary. Earth movements came to a climax in the mid-Tertiary, although no great geosyncline directly affected this country: the Tethys lay over 500 km to the south. This phase covered a period of 180 million years, and was completed by erosion which is still continuing. The relief, speed of erosion and dry land must all be much

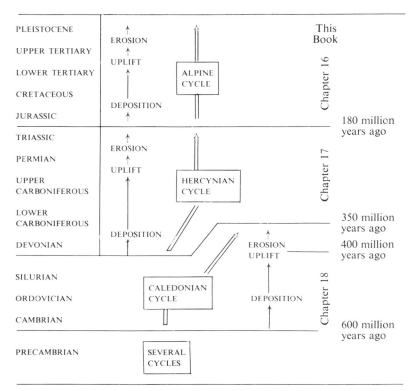

Figure 15.4 British orogenic cycles related to the geological column.

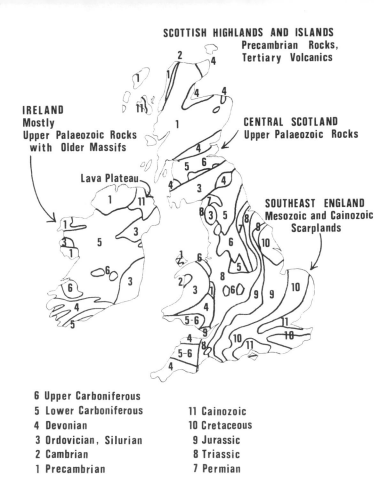

SCOTTISH HIGHLANDS AND ISLANDS
Precambrian Rocks,
Tertiary Volcanics

IRELAND
Mostly
Upper Palaeozoic Rocks
with Older Massifs

Lava Plateau

CENTRAL SCOTLAND
Upper Palaeozoic Rocks

SOUTHEAST ENGLAND
Mesozoic and Cainozoic
Scarplands

6 Upper Carboniferous
5 Lower Carboniferous
4 Devonian
3 Ordovician, Silurian
2 Cambrian
1 Precambrian

11 Cainozoic
10 Cretaceous
9 Jurassic
8 Triassic
7 Permian

Figure 15.5 A generalised geology of the British Isles. The outcrops of the major groupings of rocks are shown.

PLEISTOCENE	LAND: glacial, fluvial
UPPER TERTIARY	(mostly land) SHALLOW SEA
LOWER TERTIARY	SHALLOW SHELF SEAS; VOLCANIC PROVINCE in NW
CRETACEOUS	SHELF SEA (clays, sands, limestones); LAKE
JURASSIC	SHELF SEA (clays, sands, limestones, ironstones); DELTA
TRIASSIC	LAND: eolian, fluvial, lacustrine
PERMIAN	LAND: eolian, fluvial; SHELF SEA: limestones, sands, evaporites
UPPER CARBONIFEROUS	DELTA; SHELF SEA; GEOSYNCLINE in south
LOWER CARBONIFEROUS	SHELF SEA (limestones, reefs) DELTA; GEOSYNCLINE in south
DEVONIAN	SHELF SEA in south; LAND: fluvial, lacustrine; VOLCANIC
SILURIAN	GEOSYNCLINE; SHELF SEA
ORDOVICIAN	GEOSYNCLINE; SHELF SEA; VOLCANIC ACTIVITY
CAMBRIAN	GEOSYNCLINE; SHELF SEA
PRECAMBRIAN	A variety of GEOSYNCLINAL, SHELF SEA, LAND and VOLCANIC CONDITIONS

Figure 15.6 British rocks and their environments. Some examples of the different environments present in the British area during the past, as shown by the rock evidence.

greater today than has been normal in the geological past, when widespread seas were more common than marked continental relief.

There were other such cycles before the Cambrian period, and it will be one of our final tasks to see how these are recorded in the contorted and altered rocks of the Scottish Highlands: at least two, and probably more, can be discerned. We would expect this in eras covering over 2000 million years of time.

16 The Evolution of the British Isles: the Most Recent Orogenic Cycle

We shall now illustrate the working out of these principles of stratigraphy in the study of the geological evolution of the British Isles. Many of the principles were first suggested by a study of the Jurassic rocks and those of later date, and these rocks illustrate the principles most simply: we shall, therefore, start with them.

It has been traditional to study the oldest and most complex rocks first in a stratigraphy course. This approach has the virtue that one studies each group of rocks in the order in which they were formed, and thus follows the sequence of events through from start to finish. The main drawback with such a course is that it begins with rocks which are most likely to be poorly exposed and highly deformed, and so do not illustrate the simple application of stratigraphical principles, and which outcrop only in the remoter parts of Britain. When the student begins his course with the Precambrian and Lower Palaeozoic rocks, the later study of the Mesozoic and Cainozoic rocks seems rather tame: it is more appropriate to begin a stratigraphy course with the interpretation of the simpler situations and to see at a later date what happens when we attempt to apply the same principles to more complex situations.

To the south and east of a line running across England from the Tees estuary to the mouth of the River Exe there is a series of relatively thin layers of gently tilted rocks nearly all formed in the shallow seas which covered a varying proportion of Britain for most of the time from the beginning of the Jurassic until the Oligocene. The tilting of these rocks has resulted in the formation of a characteristic landscape pattern of alternating escarpments and vales.

Whilst these rocks were being formed, what is now the British Isles lay in a shelf sea far from the centre of orogenic activity, and the nearest geosynclinal area was the Tethys Ocean across the southern margins of present-day Europe. There was considerable variety of conditions within the shelf seas, which changed from place to place and from time to time. The rocks of north-east England were formed on an erosion surface sloping to the east, and those of the English Midlands and London area on a very shallow area which occasionally became land—the **London Platform**. There was another broad area of shallows round the Mendips, and most of Wales and Scotland were probably hilly areas supplying debris to the seas. Between these areas were zones of subsidence, in which much thicker successions of rock accumulated. In the extreme south borings have proved over 3000 m of sediment of this age in the Weald–Wessex area, and there is a similar line of troughs stretching from the Lower Severn Valley to the Cheshire Plain and Solway Firth areas. This second group, however, is largely filled with deposits of Permian and Triassic age.

Lower Jurassic (or Liassic) Rocks: Correlation by Ammonites

After a long phase of largely land conditions in the Permian and Triassic periods, a shallow branch of the Tethys Ocean spread northwards and deepened to inundate most of the British Isles, which had been worn down to a very low-lying landscape. Muddy waters covered the south and east and extended in shallow gulfs around the coasts of Scotland and Wales: these areas probably stood out as land, supplying fine debris to the seas. Although clays and thin bands of limestone alternate on the Dorset coast, most of the other Liassic rocks are dark shales. They contain a wealth of fossils, especially ammonites and reptile remains, but also those of other animals which liked muddy sea conditions, such as oysters and belemnites. The only important variation is found in the Midlands and north, where there are layers of ironstone, rich enough to be worked at such places as Scunthorpe (Lower Lias), Banbury and the Cleveland Hills (Middle Lias), and forming a low escarpment.

The Liassic rocks are very well-zoned by ammonites, and this fact enables us to reach several important conclusions concerning the conditions of sedimentation at this time. It is common, for instance, for

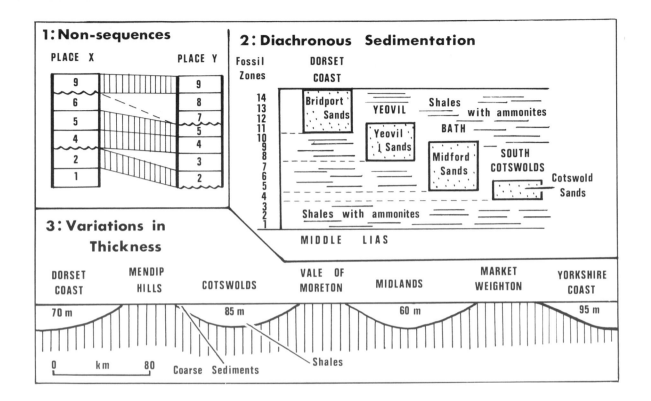

Figure 16.1 Features of the Upper Lias. In (1) the numbers 1–9 indicate fossil zones; the layers of rock follow on without any visible unconformities, but can you see where deposition did not occur? How many zones do the two areas X and Y have in common? (2) shows how the sandy sediments became younger in age as the environment in which they were formed moved southwards. (3) illustrates the effect of varying depths of water, and of degrees of subsidence, on the thicknesses of Upper Liassic rocks.

some of the fossil zones to be missing from the succession in any one area, as Figure 16.1.1 demonstrates. Two reasons have been suggested for these **non-sequences** in the rocks: either there was no deposition at the time, or the particular type of ammonite did not reach the area from which it is missing. Another important fact is that the volume of sediment deposited varied from place to place, for 2 m of shale in Dorset contains the fossils of six distinct zones, each of which normally extends through several metres of rock thickness. This is known as a **condensed sequence**. A third fact is demonstrated by the Upper Liassic rocks of southern England, where a group of sandy rocks contain the fossils of younger and younger zones as they are traced southwards (Figure 16.1.2). The special conditions of deposition in which the sands were laid down gradually moved to the south, so that the layers of rock were formed across the time planes, and are said to be **diachronous**. Finally, the uneven nature of the sea floor at this time is shown by the thicknesses of Upper Liassic rocks found in different parts of the outcrop: Figure 16.1.3 shows this. The dominant shales give way to coarser, shallow-water sediments at the thinnest parts of the succession.

Middle Jurassic: Facies Changes in Space

The Middle Jurassic rocks of England illustrate the differences in rock-forming conditions which were present at the same time. We use the word **facies** to describe particular features and characteristics of a sedimentary rock: it includes the sum of its composition, internal structures and overall geometry (ie whether it is flat-bedded, lenticular or wedge-shaped), together with the fossils it contains. Each sedimentary facies is the product of an ancient depositional environment, and the names given to facies may indicate this—eg deltaic facies, reef facies. Alternatively facies are named after descriptive aspects of the rock—eg oolitic facies, graptolitic or shelly facies. In many ways it is better to use descriptive terms, since different depositional environments may give rise to similar features, and interpretations change: thus many rock groups described originally as estuarine facies are now reinterpreted as of deltaic origin.

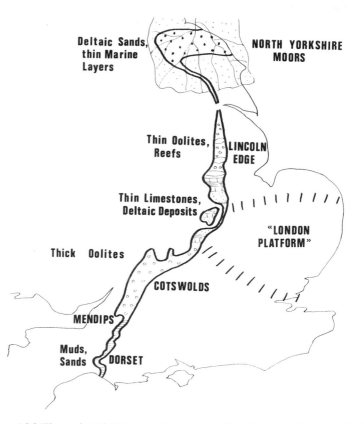

Figure 16.2 The main Middle Jurassic outcrop. Note the changing rock facies across country.

Figure 16.2 summarises the varying facies along the length of the main Middle Jurassic outcrop in England. Today's most striking relief feature in rocks of Jurassic age is provided by the thick oolitic limestones of the Cotswold Hills in Gloucestershire. Here the layers of

231

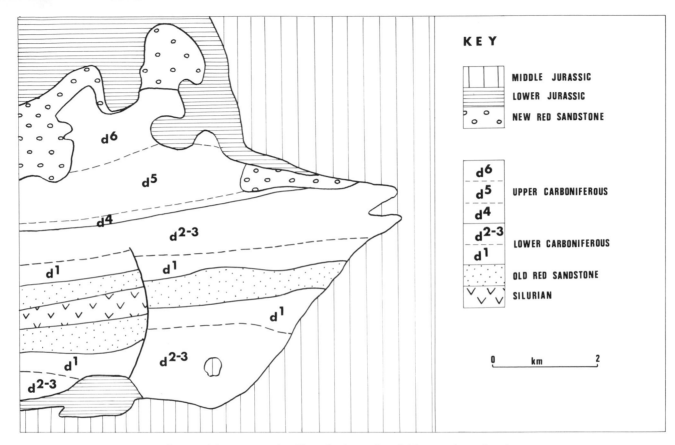

Figure 16.3 Map exercise. How do the rocks of this area show that there was a period of folding between the deposition of the Palaeozoic formations and the Mesozoic rocks? Suggest evidence to show that the Middle Jurassic seas extended farther than those of the Lower Jurassic, leading to an overlap situation. (Sketch-map based on the Geological Survey 1 : 63 360 Frome Sheet. *Crown Copyright Reserved.*)

limestone are interrupted by thin beds of marl and fine lagoonal limestones like the Stonesfield 'slate'. The best oolites are used for building stones, and the 'slate' for roofing tiles. All these rocks were formed in shallow, lime-rich seas. To the south of the shallow Mendip area, however, rocks of the same age are clays, and to the north of the shallow Vale of Moreton zone the escarpment again becomes less prominent as the proportion of clays increases. In parts of Northamptonshire the deposits are estuarine, containing fossils of freshwater shell creatures and plants, and there is another thick layer of ironstone round Corby and Kettering. This is the most valuable British source of iron ore, and is mined by open-cast methods (Figure 8.12). The Lincoln Edge area was more like the Cotswolds, being formed of false-bedded oolitic limestone and coral reefs, but north of the Humber estuary there is the greatest contrast of all. The North Yorkshire Moors, where Middle Jurassic rocks form so many of the hilltops, were the site of a delta, very similar to that in which the Millstone Grit of Carboniferous age was formed. Coarse, pale-coloured sandstones containing thin bands of plant remains and even coal seams are interrupted at three levels by marine incursions across the deltaic flats. These rocks are very similar to the Middle Jurassic rocks of Scotland, where the east and west coasts and the islands locally have sediments of this age. Some are marine and include clays, sandstones and limestones, but others are deltaic, estuarine or fluviatile. There is even a workable coal seam at Brora on the east coast of Sutherland.

The Upper Jurassic and Lower Cretaceous: Facies Changes in Time

Sedimentary facies not only change from one place to another during the same phase of geological time but are also found varying from one phase of geological time to the next in the same place. Figure 16.4 illustrates changes which can be traced in the uppermost Jurassic and the Cretaceous rocks of the Dorset coast: summarise the situation in terms of uplift in the late Jurassic and of rising sea-level during the Cretaceous—or of a regressing and transgressing sea.

The **later Jurassic** rocks saw another return to widespread marine

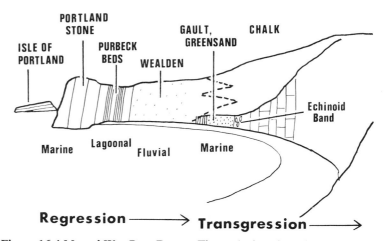

Figure 16.4 Man o' War Bay, Dorset. The rocks here have been raised into a vertical position by earth movements: the sequence can be examined easily and rapidly.

conditions, and muddy waters replaced the clear seas. On top of a thin layer of marine limestone or calcareous sand, known as the Cornbrash, the bluish Oxford Clay forms low-lying country and provides the raw materials for the largest brickmaking industries in the country round Peterborough and Bedford. The muddy conditions put an end to the clear limestone seas for a while, but these later re-established themselves to the south of Oxford, where a bed of Corallian Limestone forms a low escarpment. North of Oxford, however, the muddy conditions predominated, and the Ampthill Clay was deposited, adding to the width of the clay vale leading north-eastwards to the Wash. In Yorkshire the Corallian Limestones are at their thickest, and form the steep escarpment of the Hambleton Hills along the western edge of the North Yorkshire Moors. A final extension of the muddy waters led to the formation of the Kimmeridge Clay all over south and eastern England. It is very thin in Bedfordshire and round the Humber estuary, but thicker in Dorset and beneath the Fens and Vale of Pickering (Yorkshire), reflecting the fact that the zones of uplift and subsidence (Figure 16.1.3) were still active. The seas still impinged on the coasts of Scotland, and fossils of Kimmeridge age have been found in rocks that resemble the Old Red Sandstone on the east Sutherland coast. These remarkable rocks contain huge boulders up to 50 by 30 by 10 m in size, and it is thought that they must have been tumbled down from cliffs by something special like tsunami waves generated by earthquake shocks.

But that was the end of the Upper Jurassic marine transgression, and the highest layers in the succession show us that the sea became shallower, was restricted to a much smaller area and finally excluded. The calcareous sands and oolitic limestone characteristic of the Portland Stone are only found between the Dorset coast and Oxford. This is another of our most important building stones. The overlying Purbeck Beds are freshwater limestones containing pond-snail fossils, petrified forests and 'dirt beds' (fossil soils). Recent borings in Kent have shown that both the Portland and Purbeck rocks are very thick there.

It is evident that at this time an area of sea had been cut off from the main shelf seas covering Europe, and had been largely filled in with sediment, leaving swampy conditions in the extreme south of Britain, whilst the north must have experienced some uplift and erosion. These conditions were continued into the early stages of the **Cretaceous period**, for the area to the south of London was a zone of shallow lakes and lagoons crowned by the flat-surfaced deltas of rivers bringing debris from the London Platform area of land. The sands and clays, with occasional nodules of limestone and ironstone, bear a record of sun-cracks and ripple-marks on their bedding planes and contain fossils of giant reptiles and tiny water-fleas. Coarse pebbly beds in these Wealden Series rocks enable some of the ancient rivers to be traced, and the succession of sediments show a regular rhythm as the level of the Wealden lake rose and fell (Figure 16.5).

These rocks, which are now exposed in the central Weald of Kent, Surrey and Sussex, are covered in this area by the Lower Greensand,

IV { Middle and Upper Weald Clay (lacustrine)
     Horsham Stone (thin, sandy deltaic deposit)

III { Lower Weald Clay (lacustrine)
      Upper Tunbridge Wells Sand (deltaic)

II { Grinstead Clay (lacustrine)
     Lower Tunbridge Wells Sand (deltaic)

I { Wadhurst Clay (lacustrine)
    Ashdown Sand (deltaic)

Repetitions I and II can be broken down into a more detailed succession, which demonstrates the fluctuation of delta and lake.

6. Thick siltstones, silty clays.........Lake gets shallower

5. Thick, dark clays, with ostracods

4. Thin clays with shell beds.........Lake deepens

3. Soil bed with rootlets reaching { Shoreline
   down into sands and clays   { retreats as lake
                                     { level rises

2. Pebble-bed

1. Thick sandstones .... Deltaic deposits,
                                    falling lake level

Figure 16.5. Notice how the changing deposits reflect the deepening of the lake waters, or the advance of the delta.

which was formed as shallow seas once again swept across the whole area and buried the deltaic and lacustrine deposits. The Greensand is formed of soft, unconsolidated brown or orange sands, sometimes with hard, cherty bands, and locally has a calcareous cement. The fact that it is seldom green is due to the chemical alteration of the green iron silicate, glauconite, to the brown iron oxide as it is weathered. The presence of glauconite, however, tells us that these rocks were marine deposits. The varying importance of the different types of rock accounts for the fluctuating height of the escarpment which is normally associated with the outcrop of the Lower Greensand. It forms Leith Hill (320 m O.D., the highest point in south-eastern England),

Worbarrow Bay, Dorset. The rocks are dipping northwards: the oldest are the Upper Jurassic rocks on the extreme right; the central ridge is Upper Cretaceous chalk, and behind that are lowlands on the Tertiary sands. Relate this succession of rocks to that in Figure 16.4. (*Aerofilms*)

but at times disappears altogether as a feature in the landscape. The overlying Gault Clay and Upper Greensand show that the sea extended to the west of England and then was filled in by coarse sandy deposits (Figure 16.6).

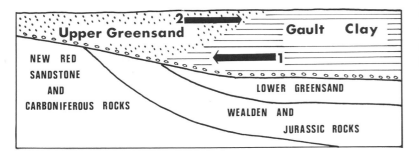

| EAST DEVON | ISLE OF PURBECK | ISLE OF WIGHT | EAST KENT |

Figure 16.6 Upper Greensand and Gault Clay. These two types of rock were formed at the same time by a transgressing sea (1) advancing over the tilted and eroded older rocks; and by a succeeding period of deposition filling the seas with coarse debris from the west and causing it to retreat (2).

At the same time north-eastern England, on the far side of the London Platform, was subjected to quite different conditions as an arm of the sea covering Germany and Russia swept in. Thick clays containing numerous ammonites just like those found in central Europe, but different from those in the Weald and northern France, were formed. They are topped by the unusual deposit of Red Chalk.

The Chalk: Unique Sedimentary Facies
The final episode in the Jurassic–Cretaceous phase of quiet, shallow-sea conditions was the most remarkable. It saw perhaps the most complete coverage of these islands by the sea that they have ever known. Some geologists believe that the whole area, including the mountainous north and west, was inundated by the widespread seas, in which the unique deposit of **Chalk** was formed to thicknesses reaching almost 500 m. Chalk is an unusually pure calcium carbonate rock, formed largely of microscopic plant and animal debris together with scattered larger fossil shells and their broken debris. Very little sand or mud could have been supplied to this sea, and although the lower layers contain up to half their volume of clay and have a greyish colour the greater majority of the Chalk is dazzling white in colour with less than 1 per cent of non-calcareous material. The upper layers contain many siliceous nodules and flat masses of flint, which occur along the bedding planes but sometimes also across them (Figure 16.7). This arrangement suggests that they were formed after the

Figure 16.7 Flints in the chalk at Flamborough Head, Yorkshire. What evidence suggests that the flints were formed after the chalk?

Chalk, probably as it was drying out and water was percolating through the rock. The source of so much silica, however, is a great problem. Some geologists have suggested it originated from the solution of sponge skeletons, and others have put forward the idea that the silica came from the volcanic activity associated with the early stages of the opening up of the North Atlantic Ocean.

The Chalk seas were therefore the scene of slow limestone deposition, and it has been estimated that the rate of accumulation might have been something like 1 cm every 1000 years if it was anything like the similar deep oceanic oozes of today. This resemblance in characteristics led geologists to classify Chalk as a deep-sea deposit, but harder, nodular bands like the Chalk and Melbourne Rocks and Totternhoe Stone are definitely shallow-water sediments, and many of the common sea-urchin and mollusc fossils are never found in waters deeper than 200 m.

The formation of the Chalk ended the Mesozoic sedimentation, and it was followed by a break and considerable uplift during which a lot of Chalk was removed by erosion. This was the herald of the Alpine earth movements which were beginning to raise up mountains far to the south, but their main effects were not to be felt for some time in Britain.

Life in Jurassic and Cretaceous Times

These two periods have been called **'The Age of Reptiles'**, for it was at this time that that group of animals dominated the scene. On land the dinosaurs multiplied into many different varieties, including small, fleet-footed types and great lumbering monsters. The largest Brontosaurus was 30 m long and probably weighed 35 tonnes: it had to live in swampy conditions which helped to buoy up its weighty body which its legs could not otherwise have supported and to reduce its diet of plant leaves to an edible pulp. It was a herbivore (plant-eater), but some of the normally more active carnivores (meat-eaters) like Tyrannosaurus grew to 16 m long. Some reptile groups lived in the seas, and the ichthyosaurs and plesiosaurs were up to 10 m and 20 m long respectively. Pterosaurs developed web-like wings which were used for gliding, soaring and eventually for proper flying. The warm-blooded, feathered birds also arose during the Jurassic period and probably had a reptile ancestor. At the end of the Cretaceous, however, most of the important reptile groups became extinct. It is difficult to see the reason for this, but it may have been connected with the physical conditions such as the spread of the Chalk seas—though these did not cover the whole world—or with the changing vegetation.

The land plants had changed little during the Mesozoic era. The coniferous plants, which had first been found in the Coal Measure forests of Carboniferous age, had developed with their close relations the ginkgos and cycads, and with many ferns. Towards the end of the Cretaceous the first fossils of flowering plants are found in the rocks, and this group soon became the most important of all. Fossils of pollinating insects are also found from this time onwards, but it is uncertain which appeared first. Perhaps the large, herbivorous reptiles had become too specialised in their diet and could not readapt themselves to these new plants: the carnivores would then be deprived of their staple diet.

The Jurassic and Cretaceous systems were also particularly important for the development of the invertebrate animals—insignificant no doubt when they were alive, but of the greatest importance to geologists today. Just as this is the 'Age of Reptiles' in one sense, so it is **'The Age of Ammonites'** in another. These animals living in coiled shells spread northwards from the Tethys Ocean, with one family succeeding another: each had new features which had evolved in the deep geosynclinal trough across southern Europe. They provide us with some of the best-established correlations in the whole succession of rocks formed in past ages, although they tended to avoid seas rich in lime. The Lower Jurassic rocks and the Oxford and Kimmeridge Clays contain plenty of fossil ammonites, but there are few in the oolitic limestones and in the Chalk. The other mollusc groups were also extremely important, and the sea urchins had reached a high level of development by the time the Chalk seas arrived: they are used as zone fossils for certain Chalk horizons (Figure 9.21). Whilst the ammonites are used for zoning the Jurassic and Lower Cretaceous, a variety of fossils—sea urchins, brachiopods, free-swimming crinoids and belemnites—are used in the Chalk.

Some of the fossils that had been important in the Palaeozoic were only represented in the Mesozoic by their most advanced species, such as the terebratulid and rhynchonellid brachiopods, and the *Pentacrinus* crinoids with their star-shaped ossicles. The corals were completely changed, and nearly all the fossils of this age belong to the scleractinian group.

At the end of the Cretaceous many groups of animals became

extinct besides the reptiles. These included the ammonites and the belemnites.

The Lower Tertiary: Marine and Continental Conditions

There is a complete break in Britain between the Tertiary rocks and the underlying Chalk, although the unconformity which was produced is nowhere very striking. The northern and western parts of Britain were also affected by these movements, which led to the initiation of major river systems draining towards the east. South-eastern England was not uplifted, but two areas were warped down and were drowned by part of a sea that extended across to the Paris Basin and northern Belgium. These two areas, the Hampshire and London Basins, were separated by an island due to the updoming of the Weald, and they have somewhat different Tertiary sequences. Both record the fact that these basins occasionally subsided, making room for the sea to advance over the top of sediments that had been filling in the previous depth of water. The rocks of the London Basin only testify to two such cycles, whereas the Hampshire–Isle of Wight area has five, extending over a longer period of time. Whereas the London Basin rocks are all **Eocene** in age, those round the Solent estuary are Eocene and **Oligocene**. Each cycle began with the formation of pebble beds, marine clays or sands, containing fossils of fish and marine molluscs, and ended with the extension from the west of coarse deposits with evidence of shallower water and even continental conditions of sedimentation. Figure 16.8 contains the details. The last sediments preserved are early Oligocene, and it seems probable that by the middle of the Oligocene period all formation of new sediments had ceased in south-eastern England.

Close study of these rocks in areas like the Isle of Wight shows up some of the differences between rocks which were formed on the land and those formed in the sea.

1 **Rocks Formed on the Land.** These include river-laid sands and pebbles; freshwater lake clays and limestones which often include the distinctive, thin-shelled snail shells; plant and leaf beds which do not

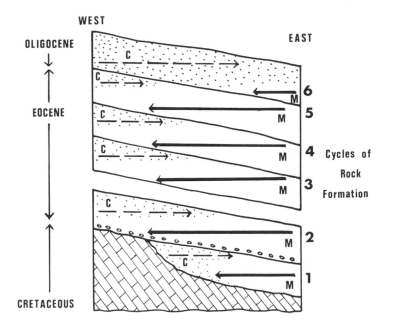

Figure 16.8 Lower Tertiary sediments. The rocks can be divided into 6 cycles of sedimentation, all following a similar pattern as the seas advanced from the east (thickened arrows) bringing marine conditions (M). They then retreated (broken arrows) as they were filled by sediment and gave way to continental conditions (C). Only the first 2 cycles are represented in the London Basin, but all are found in the Hampshire Basin and Isle of Wight. Compare this diagram with Figure 16.6. How can we tell that a rock had a marine, or continental, origin? What would be the difference between the rock successions at the eastern and western ends of the area affected?

show too much disturbance or other indications of transport. Except for the lacustrine clays and the plant beds, land deposits seldom contain fossils. Bones of land mammals are extremely rare.

2 Rocks Formed Beneath the Sea. These are characterised by the fossils of marine animals: we know, for instance, that certain types of gastropod and lamellibranch shells are only found in sea water today, and many of these had close relatives in the early Tertiary. Rock-types of marine origin vary from clays to sands, and those formed close to the shore may contain drifted and broken plant remains.

Discuss the significance of these characteristics, and add to the list as you study rocks formed at other periods of geological time.

Volcanic Activity in the North-west

Other events were taking place at the same time in the extreme north-west of Britain. As the south-east was subject to occasional uplift and subsidence, so the north-west of Scotland was part of a great belt of **volcanic activity** extending from Iceland into Ireland as the North Atlantic split open. Lava flows were poured out, building up tremendous volumes of basalt rock in such places as Northern Ireland (eg Giant's Causeway) and Mull (over 2000 m thick). Such thicknesses required many eruptions, as the individual flows were never more than 30 m thick. Between the eruptions the lava surface was weathered, and soils were formed. In some of these soils sub-tropical plants of Eocene age have been found as fossils on Mull, and Oligocene plants occur in clays formed on top of the Antrim basalts. Volcanoes dominated what are now the islands of Skye, Mull and Arran, but today they are only represented by their deeply eroded stumps. These volcanoes often seem to have collapsed at the end of their active life, carrying down fragments of layers of rock that have otherwise disappeared from the area. Such cauldron subsidence left a complicated pattern of volcanic and intrusive igneous rocks in these areas. Intrusive masses of granite and gabbro are also found in western Scotland and Northern Ireland. Radiometric dates for the granites are all around 50 million years—ie Eocene in age. The island of Arran was widened by over two km in the course of dyke intrusion and an almost parallel group of dykes spread across southern Scotland and into northern England: the Cleveland Dyke (Figure 10.6) is the best known of these.

VOLCANIC CENTRES

LAVA PLATEAUS

Figure 16.9 Tertiary volcanic activity in western Scotland and northern Ireland.

Uplift in the Middle Tertiary

There are no rocks in the British Isles which can be assigned with certainty to the Miocene period. This was the time when the effects of the great Alpine earth storm reached these islands. Nothing happened in Britain to parallel the upheavals which produced the Alps and Pyrenees, but an indelible impression was left behind. To the south of the Thames there were thick layers of soft Jurassic, Cretaceous and early Tertiary sediments. These were folded into short, east–west folds, plunging at each end and only a few miles long: some are shown on the map in Figure 5.10. Farther north these younger sediments were not so thick, and the resistant, ancient rocks on which they rested yielded to the pressures in a series of faults. The whole area was tilted up to the north and west once again, giving rise to the distribution of relief we find today, and helping to expose the oldest rocks in these most heavily eroded areas.

The Upper Tertiary and Pleistocene: Mainly Land

The most recent rocks in Britain, apart from the thin mantle of drift deposits left by ice, wind and river action, are the deposits of East Anglia, for long known as the Crags: this was an area which was warped down towards the North Sea during the Pliocene and Pleistocene whilst most of Britain was land. The rest of the country was still undergoing erosion, which left no tangible results in the form of rocks and fossils. The oldest of these Crag deposits is the Coralline Crag, formed of colonial polyzoan fossils and lamellibranchs embedded in limy sands. The faunas in the later Crag deposits, now included in the Pleistocene, indicate that the climate was getting cooler. We are probably still in the middle of the Ice Age which began in the Pleistocene period, for the glaciers have advanced and retreated several times, and at the present moment we are experiencing the later stages of ice retreat. We cannot tell whether the ice will eventually advance again, but, if it does, it will not do so for many thousands of years.

The fossils in the Crag deposits of East Anglia record the drop in temperature at the end of the Pliocene and the beginning of the Pleistocene period. They have thickened shells, and include varieties which are found today off the coast of northern Norway. The ice advanced from the highlands of this country and Scandinavia to cover much of northern Europe. At its greatest extent it reached the Bristol Channel–Thames line (Figure 16.10). The fluctuations of the position of the ice-sheet front can be traced by studying the layers of boulder clay and the morainic deposits formed at the time. The succession of pollens in peat deposits, and the variation of sea-level as recorded in the terraces of river valleys like the Thames, tell us how the climate warmed up or got cooler. At least three periods of glacial advance affected Britain, and it seems that the last of these left a complicated

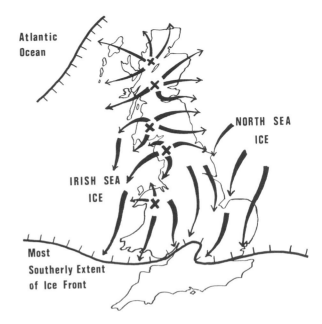

Figure 16.10 The Pleistocene ice sheet. This map shows the southern limit of ice advance, and the centres of ice dispersal—marked by an X. Can you name these centres?

pattern of retreat deposits with occasional short-lived readvances. The warmer interglacial periods between saw higher sea-levels, and one at least brought a warmer climate to Britain than that which we enjoy today. Some of the main events are summarised in Figure 16.11.

This Ice Age brought many changes to the relief of Britain, especially in the highland areas where erosion was most intense, and thin deposits were spread over the lower-lying areas. The actual relief forms produced and the deposits have been described in Chapter 7. The climate of the areas surrounding the ice-sheets was also very cold,

EVENT	ICE MOVEMENT	CHANGES IN SEA-LEVEL	DEPOSITS	LANDFORMS, DRAINAGE
8. Postglacial time	Retreat	Rise	Deep Thames channel filled in; Fens silted up	Low Raised Beach of Scotland (8 m)
7. 3rd Glaciation: Weichselian	Advance down Irish Sea and to east coasts	Fall to level below present (c. −50 m)	Marginal lakes—L. Lapworth in Shropshire, L. Pickering in Yorkshire	Retreat of ice at end broken by Scottish Readvance (High Raised Beach, 30 m) and by Highland Readvance Deep Thames channel carved
6. Last Interglacial: Ipswichian	Retreat	Rise		Avon, Severn Valley terraces; Taplow (16 m) and Flood Plain terraces of Thames
5. 2nd Glaciation: Gippingian	Advance	Fall	Second series of drift deposits over area shown in Fig. 16.10. Ice fanned out from English Midlands	
4. Great Interglacial: Hoxnian	Retreat	Rise		Boyn Hill (30 m) terrace of Thames; Goodwood Raised Beach (Sussex)
3. Earliest Glaciation in Britain: Lowestoftian	Advance	Fall	North Sea Drift and Lowestoft till in East Anglia (ie two phases of advance). Oldest drifts	Thames diverted southwards through Finchley depression, and then into present valley
2. Climate getting colder	(Advances and retreats farther north)	Fall to c. 100 m	Arctic Freshwater Bed Cromer Forest Bed Series: peats and deltaic series—slightly warmer Crags with increasing numbers of Arctic molluscs	
1. Marine transgression: Calabrian		Rise to 200 m	Red Crag	200 m Raised Beach on Downs, Chilterns, and around Wales

Figure 16.11. Use the facts summarised on this chart to make your own diagrams of (*a*) the positions of sea-level at the various stages; and (*b*) a cross-section of the Thames valley with its terraces.

241

resembling our present tundra, and the melting ice often gave rise to extensive lakes ponded back by the local relief until overflow channels could be cut (eg Figure 7.5). Freeze–thaw action caused many of our hills to be mantled with rubbly 'Head' deposits, and winds blowing across the often unvegetated surfaces resulted in fine, loamy loess being dropped along some of our valleys (where it is known as brick-earth). The weight of the ice also led to vertical movements of the land, and as it melted the isostatic recoil led to beaches becoming raised above the line of wave action, especially in Scotland where the ice had been thickest.

As the ice finally left Britain the climate grew warmer once again, the sea-level rose and river action took over the work of denudation. The fact that the glacial landforms are still relatively fresh in our highland areas shows that the rivers have not had enough time in the last 10 000 years or so to make many alterations. Yet many of the deep, glacial 'ribbon lakes' have been partly filled in with debris (eg the head of Ullswater in the Lake District), and some have been divided by a river building a delta of debris across the lake (eg Buttermere and Crummock Water). The most recent geological record is based on the evidence provided by the landforms rather than by the deposits.

Life in the Cainozoic Era

This was the beginning of 'The Age of Mammals'. After over a 100 million years of insignificant existence on the Earth an unprecedented evolutionary radiation took place, giving rise to an immense variety of species from the earliest Eocene. At first primitive representatives of the major mammal groups often grew to enormous sizes, whilst others, like the horse, still remained small, fox-like creatures. We can trace the development of many of the groups we know today (eg Figure 9.27) in some detail, and relate them to important changes in the habits of the animals concerned. By the end of the Tertiary the horses, rhinos, wild cattle, sheep and pigs, elephants, cats, dogs and bears were all looking very similar to their modern successors. The primitive platypus, kangaroo and opossum had been isolated in the island continents of Australia and South America.

When one examines the fossils of British Tertiary rocks, one is impressed by the abundance of three groups in particular. Two of them are molluscs, the lamellibranchs and gastropods, and dozens of varieties can be obtained from most exposures of Tertiary rock. Neither of these is very helpful to the geologist attempting to zone the rocks and correlate them with others in a distant area. The third group is more important to such a person, and is very valuable to the oil geologist. It is the Foraminifera, a group of protozoans—fossils of animals which evolved rapidly and became widespread in distribution as part of the plankton of the time. Most of them are microfossils, but some are up to 2 cm across.

The most interesting fossil evidence in the Pleistocene rocks relates to **the advent of human beings**. The earliest man-like creatures seem to have lived in the warmer lands of Africa, but spread to most parts of the world in the interglacial periods. At first there were several species of primitive man, but during the final advance of the ice all but Homo sapiens became extinct. The fossil evidence includes skeletons (usually very fragmentary) and flint tools, the latter showing how man's mind was developing so that he was able to become more and more independent of the natural environment and his own bodily limitations.

The fossil skeleton of man shows very few differences from that of other animals belonging to the primate group, such as the ape or monkey. It seems that man and the other primates must have had a common ancestor. The main changes are twofold—to allow man to stand erect, and the enlargement of his brain, especially in the vital frontal portions. The latter gave him an outstanding advantage by enabling him to consider the future in the light of past experience, and to think out rational answers to problems. It is not clear when the decisive characteristics which divide man from the other animals were acquired, but by the later part of the Old Stone Age (approximately 35 000 years ago) he was producing most wonderful works of art with a clear religious significance in the caves of southern France.

17 The Evolution of the British Isles: the Upper Palaeozoic Orogenic Cycle

The orogenic cycle which preceded the Alpine began with deposition in southern England in the Devonian, as Figure 15.4 shows, and this phase of sedimentation spread to the whole country during the early Carboniferous period. At the end of the Carboniferous uplift took place, and the Permian and Triassic rocks record a largely land environment, typical of the post-orogenic stage.

The rocks formed at this time have an outcrop which includes the South-west Peninsula (Devon and Cornwall), south Wales, the English Midlands, the Pennines and north-east England and parts of the Midland Valley of Scotland.

The Upper Palaeozoic Geosyncline

During the Devonian period only the southernmost part of the British Isles was covered by the sea (Figure 17.1). Only the southern tip of Ireland and a strip extending eastwards from the South-west Peninsula were in this zone; elsewhere there were mountains, raised at the end of the preceding Silurian period, or basins of continental deposition, giving rise to the post-orogenic deposits known as the Old Red Sandstone (see Chapter 18).

The marine rocks were formed in a deepening sea, which was part of an extensive trough of subsidence extending across central Europe at this time: the rocks of Brittany, the Ardennes, the Rhine Slate Plateau and the Harz Mountains were also formed in the same general environment. The sequence of rocks in Devon is interesting because it is typical of so many areas which have later been involved in the formation of folded mountains.

The oldest Devonian rocks are known as the Dartmouth Slates and

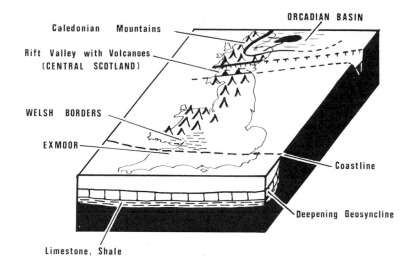

Figure 17.1 Devonian conditions. Notice the extent of the sea covering Britain at this period, and the situation of the main areas of deposition.

are reddish slates and sands containing some plant and fish debris like the Old Red Sandstones farther north. These are covered by shallow water marine sands and shales, and the Middle Devonian rocks include local lenses of volcanic rocks and massive limestones which contain reef-building fossils. The Upper Devonian rocks of south Devon and most of the Devonian age rocks in Cornwall are mainly fine shales which contain the fossils of creatures which swam in deeper waters (eg goniatites), but few bottom-living forms. This record suggests that the seas were gradually getting deeper. In north Devon these rocks are much thicker and are interleaved with red sands and silts derived from the continental area to the north: the coast was near at hand.

The Carboniferous rocks of Devon have been called the Culm but are equivalent in age to both the Lower and Upper Carboniferous else-

243

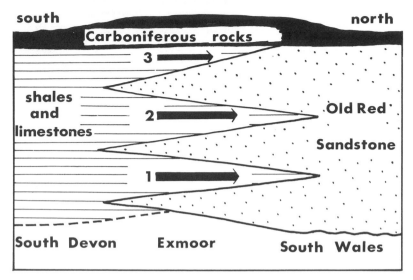

All younger rocks rest on top with marked unconformity		Folding and uplift
UPPER CARBONIFEROUS	Flysch facies: alternate shales and greywackes	Seas filled in by rapid sedimentation after local uplift
LOWER CARBONIFEROUS	Deep-sea oozes	Seas deepening with slow rates of sedimentation
DEVONIAN	Deep-sea shales	
	Shallow sea muds, sands, limestones	
	Old Red Sandstone facies	Land conditions, or just offshore

Figure 17.3 Upper Palaeozoic rocks in south-west England and an interpretation of their origin.

Figure 17.2 The Devonian shoreline. The rocks of Exmoor show how the sea advanced northwards three times to deposit typical marine rocks, and how the three sandstones were formed in the intervals, on the land.

Carboniferous an uplift led to more rapid sedimentation of some thousands of metres of flysch facies. At times deltas extended southwards into the area from Wales, and this sequence of events ended in the Upper Carboniferous when the whole area was uplifted to form mountain ranges.

The Lower Carboniferous: A Variety of Facies

By the end of the Devonian period the Caledonian Mountains had been reduced by erosion to a large extent, and the southern seas advanced northwards. The rocks of Lower Carboniferous age record this advance, together with the deepening of the geosyncline in the south, and then the Upper Carboniferous witnessed the filling in of both areas of sea and the raising of new mountain chains. The earlier part of the Carboniferous period, then, was a time of marine transgression across the worn-down remnants of the Caledonian Mountains and the Old Red Sandstone deposits. Conditions on the sea floor varied and so the types of rock formed at the time record a number of distinctive changes from place to place (Figure 17.4). Some areas had a gently subsiding sea floor; in others there was more sudden subsidence and

where. The Lower Carboniferous horizons are thin, dark, lime-rich rocks and cherts including radiolaria. These rocks may have been formed in deep waters and include volcanic layers. The Upper Carboniferous rocks record a dramatic change. They are a monotonous succession of extremely thick interbedded shales and sandstones. The sandstones are often of a greywacke composition, may have graded bedding, and the bottom surfaces show structures associated with turbidites, like load casts and flute marks. This association is very much like the flysch of the Alps. Figure 17.3 summarises the progression.

An interpretation of this succession of rocks suggests that the Devonian and Lower Carboniferous saw the extension and deepening of the area of marine deposition, but that at the beginning of the Upper

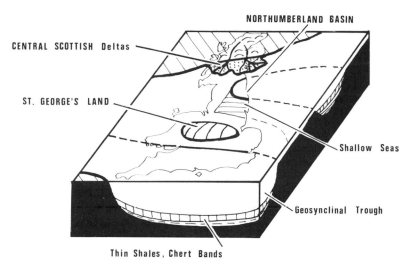

CENTRAL SCOTTISH Deltas

ST. GEORGE'S LAND

NORTHUMBERLAND BASIN

Shallow Seas

Geosynclinal Trough

Thin Shales, Chert Bands

Figure 17.4 Lower Carboniferous seas. The Upper Palaeozoic seas had their greatest extent at this time.

more rapid sedimentation; and in central Scotland, at the farthest extent of the transgression, most of the sediments were fluviatile or deltaic. The Lower Carboniferous facies are amongst the most varied of any geological period.

The best known rock-type of the Lower Carboniferous is the massive, well-bedded grey bioclastic limestone, often referred to as the 'Standard' or 'Mountain' Limestone. In the Bristol area, and in south Wales, it forms an uninterrupted succession up to 1000 m thick and contains plentiful fossils: some bands are almost entirely composed of crinoid debris, and corals and brachiopods are common. The Mendip Hills, with their caverns, dripstones, potholes and bare, dry, rocky surface, are formed of this Carboniferous Limestone, and the Avon Gorge at Bristol is cut through it.

This area bordered on the geosyncline, and has a complete record of deposition carrying on from the underlying Devonian rocks, whereas other parts of Britain usually have a definite gap between the two groups of rocks represented by an unconformity.

Central and northern England and Scotland had an amazing variety of conditions for such a relatively small area. Much of the English Midlands remained land for the whole period (an area without Lower Carboniferous rocks, known as St George's Land); the southern and central Pennine area was a deep, down-faulted basin, as was the Northumberland–Durham area: between these were shallow seas; and between central Scotland and north Northumberland there was an area of low-lying plains. Many distinctive rock facies were formed at this time.

The areas which were deep basins now have thick deposits of shale, containing thin-shelled lamellibranchs and goniatites, alternating with thin layers of black limestone. The shallower seas of those times have a variety of reef limestones: some are formed of great masses of colonial corals piled on top of each other, whilst others are built up of algal and bryozoan skeletons. Along the edges of the deeper basins a series of reefs tended to grow, and the hard, structureless limestone stands out in the present landscape as rounded knolls. The northern part of the Pennines experienced a lesser degree of subsidence and a type of cyclic deposit known as the Yoredale facies accumulated. Each cycle begins with limestone (clear, shallow off-shore conditions), followed by shale (inshore, muddy conditions giving place to estuarine and lagoonal), then by sandstone (fluviatile or deltaic) and finally by seat-earth and coal (swamps on delta top). Subsidence was intermittent and apparently sudden, because the next limestone succeeds the coal with little or no shale intervening. Marine incursions were therefore repeated, but did not persist for long. The average thickness of the cycle is about 30 m of rock, and one such succession outcrops beneath Hardrow Force in Wensleydale (Figure 5.4). Figure 17.5 shows a succession which has an outcrop in a valley tributary to Swaledale. It is thought that the repetition of the sediments occurred as distributaries swung across a delta surface in a shallow sea.

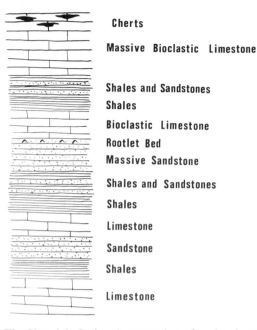

Cherts

Massive Bioclastic Limestone

Shales and Sandstones
Shales

Bioclastic Limestone
Rootlet Bed
Massive Sandstone

Shales and Sandstones

Shales

Limestone

Sandstone

Shales

Limestone

Figure 17.5 The Yoredale Series. A succession of rocks charted in a stream section tributary to Swaledale, Yorkshire, in the rock formation known as the Yoredale Series. Repetition of certain rock-types takes place, but each band in the sequence is not present every time. Construct a generalised unit to include all the possibilities.

The Yoredale facies is also found, often in modified form, in Northumberland and the Midland Valley of Scotland, where the Lower Carboniferous also included a different type of rhythmic deposit known as the Cementstones, which are locally over 300 m thick. These consist of repeated alternations of shale and argillaceous dolomite, which may have been formed annually as wet and dry

seasons followed one another. Near Edinburgh many oil-shales were formed in well-vegetated lagoons. Volcanoes were also erupting mainly olivine basalts at this time in Scotland, and many of the lava surfaces between flows have been weathered to a reddish soil. These lavas form the hills around Glasgow and many of the associated necks stick up as isolated hillocks all over the Midland Valley including Edinburgh Castle Rock and the adjacent Arthur's Seat.

The Upper Carboniferous Deltas
The Midland Valley of Scotland was the farthest north the Lower Carboniferous sea reached. There the Upper Carboniferous began with a group rather like the Coal Measures followed by a return of the Yoredale facies and then by a sequence of sandstones and fireclays. The latter are used to make refractory bricks for furnace linings. Deltas advanced southwards from the Scottish Uplands and outwards from St George's Land to cover much of central and northern England as the sea withdrew. At first many of the rocks in these areas were current-bedded gritty sandstones, often containing plant debris, and interbedded with thin marine shales in which are found fossil goniatites, indicating temporary readvances of the sea. Many of the minerals in these sandstones have been traced back to distinctive rocks in Scotland and even in Norway. Beyond these deltas marine limestones were still being formed (eg in Cumberland). Volcanic activity continued here and there in Scotland, particularly in Fife, where it is now known to have occurred in Coal Measure times.

This group of rocks, known as the **Millstone Grit** in the central Pennines where they are over 2000 m thick, forms a transition between the seas of the Lower Carboniferous and the widespread coal swamps of the Upper Carboniferous.

The deltas had rapidly spread out to cover most of the country with low-lying, swampy conditions. As mountains began to rise across central Europe more and more sediment was poured into the geosyncline, but the central and northern parts of the country were covered by these extensive deltas (Figure 17.6), beneath which there

Lower Carboniferous Cementstones, Murroch Glen, Dunbartonshire. The alternation of dolomite and soft clay shows up well on this hillside. (*Crown Copyright*)

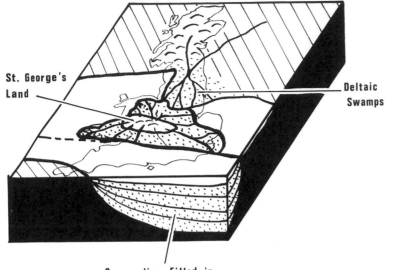

St. George's Land

Deltaic Swamps

Geosyncline Filled in

Figure 17.6 Coal Measure times. As the Coal Measure swamps filled the seas with sediment St George's Land was submerged.

was continuous subsidence as the weight of sediment increased. In places over 3000 m of rocks were deposited to form the **Coal Measures** in the manner outlined in Chapter 8.

Britain was particularly fortunate to have so much coal laid down over its surface. Coal has been the basis of our economic and political power in the recent past—though other fuels are now taking its place. The conditions of deposition meant that each local area had a unique group of coal seams with rocks between. Those in the south, including the South Wales and Kent coalfields, have two sets of coal seams separated by a thick layer of sandstone, which forms the moors between the deep Welsh mining valleys. In the English Midlands there are several small coalfields, once probably all part of the same basin of

deposition but now divided up by faults. All their valuable seams are in the lower horizons, and the upper layers of Coal Measure rock are mostly red marls formed in an arid environment. An important addition to the Midland Coal Measures is a seam of 'blackband' iron ore, which was the original basis of the metal-working industries in the area. Farther north, flanking the Pennines, are our richest coal seams. It is interesting to compare the East Pennine and Lancashire coalfields on either side of England's 'backbone': whereas the former has many thick seams which are mined easily, they often become split into several seams to the west, and the Lancashire coalfield is also broken by faults so that it is gradually declining in importance. It seems, however, that the deltas here drained into a sea to the west, and that the continued sedimentation was gradually submerging St George's Land to the south. The Northumberland and Scottish coalfields all have their productive seams in the lower part of the succession and the upper layers are once again characterised by 'barren', coal-less, red-coloured rocks. These were produced by erosion and deposition on the land, conditions which advanced slowly southwards.

The Hercynian Mountains

Deposition was interrupted at the end of the Carboniferous period by another phase of mountain-building, known as the **Armorican orogeny** (sometimes also as the Hercynian, or Variscan, in other parts of Europe). The movements had been gaining in momentum and moving northwards from central Europe, and the later stages of the Coal Measure epoch are characterised by folding and faulting in Britain. The most-affected part of Britain was the south, where the thick, soft, geosynclinal sediments were intensely folded, altered and intruded by masses of granite. Chains of mountains, running from east to west, dominated the scenery of this zone, and it is thought that they were often topped by volcanoes, such as the one that has been completely worn away from the top of Dartmoor. The granite intrusion also led to mineralisation of the surrounding rocks, where veins of tin, tungsten, zinc, lead and copper were emplaced (Figure 10.14), and hot fluids caused the margins of the granite near St Austell to be altered to china clay. The Dartmoor granite has been dated as 300–280 million years old by radioactive techniques.

Farther north, however, the effects were not so marked, and the folding in areas like the Malverns and the Pennines has a north-south trend. The folds are more open, and block-faulting was common. There was some local granite intrusion and mineralisation in these areas. As was the case with the Alpine orogeny the earth movements associated with the Armorican began earlier than the main time of uplift: unconformities are found in Lower Carboniferous rocks, and the effects lasted into the Permian period. The north–south folding is thought to be slightly earlier in time than the east–west trends.

The New Red Sandstone Deserts

Britain had become an area of dry land, crossed by chains of mountains often topped with volcanoes. The agencies of erosion set to work and produced new rocks from the broken, weathered debris. The rocks formed in the Permian and Triassic periods in Britain are nearly all associated with continental conditions, and we include them in a single group. Elsewhere in the world quite different rocks were formed at this time, and there is a fundamental difference between the fossils found in Permian and Triassic rocks, where the division between Palaeozoic and Mesozoic life is made.

When the Armorican Mountains had been raised out of the geosyncline that had existed across central Europe, only one geosyncline remained in that continent. This was **the Tethys**, far to the south of Britain, along the line of the present Mediterranean Sea. Britain had at last become a relatively stable area after all the upheavals of the Palaeozoic era. From this time onwards the country was covered by shallow seas reaching northwards from the Tethys, or by land conditions. During the Permian and Triassic periods, for instance, there were two attempts by seas to invade Britain, but neither was successful.

The **Permian rocks** of England are formed of the coarser conglomerates, breccias and sands produced by the early weathering of the newly formed mountain chains. They were deposited in basin-shaped areas cut off from rain-bearing winds in Devon, the Vale of Eden in north-west England and in the north-east. The first deposits contain fragments of Carboniferous rocks, but later erosion stripped off the rocks down to the Precambrian. We can find similar deposits forming today in low-lying arid areas close to mountains in such places as western USA and southern Siberia. Near the mountain slopes there are thick spreads of coarse gravel, then there are dune belts, and in the centre of these basins of inland drainage there are playa lakes, floored by salt layers (Figure 5.14).

A great arm of the sea, cut off from the Tethys and known in Germany as the Zechstein Sea, extended across northern Europe, and into north-eastern England. Then it began to dry up, and the increasing salinity led first to the extinction of the fishes living in the sea, and then to the deposition of the salts held in solution. A layer of Magnesian Limestone, up to 250 m thick, extends from the Durham coast through to Nottinghamshire, forming an escarpment. As the sea dried up it got smaller (Figure 17.7) and other salts were precipitated including gypsum, anhydrite, rock salt and potassium salts. The succession found in a borehole beneath Whitby in Yorkshire is shown in Figure 17.8. Many of the salt deposits of Europe, including the most famous at Stassfurt in East Germany (Figure 8.13), were formed at this time. Scottish Permian rocks are similar to the sandstones, and are largely confined to the south-west.

The **Triassic rocks** of Britain continue the picture of land being worn down in desert conditions. The threefold division of these rocks found on the continent of Europe, and implied in the name of the period, is not evident in this country, because the sea which deposited the Muschelkalk Limestone failed to reach our shores. Rocks of the older Bunter Series, and the more recent Keuper Series, are very important, and cover more of England's surface than any other single age group.

The Bunter sandstones are typical desert sands, with rounded, millet-seed grains, dune-bedding structures and faceted ventifact pebbles. A study of the dune bedding has led to the conclusion that the winds came mainly from the east at this time, and this fact, together with the

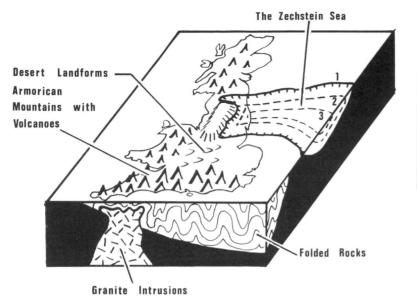

The Zechstein Sea

Desert Landforms
Armorican
Mountains with
Volcanoes

Folded Rocks

Granite Intrusions

Figure 17.7 New Red Sandstone: Permian conditions. Notice how the Zechstein Sea covered a smaller and smaller area as it dried up. Boundary 1 marks the outcrop of Magnesian Limestone; 2 the boundary of rock salt; 3 the boundary of potash salts.

The Whitby Borehole was sunk through the following Permian rocks:

A typical series of deposits resulting from the evaporation of seawater would be:

> 5. Polyhalite (potassium-magnesium-calcium sulphate)
> 4. Rock salt
> 3. Anhydrite, gypsum
> 2. Dolomite, limestone
> 1. Marl

Upper Permian Marl
   Top Anhydrite (1–1·5 m)
Salty clay (3 m)
   (3) Upper Evaporites: (65 m)
      Upper rock salt
      Potash rock (Sylvine)
      Lower rock salt
      Anhydrite
      Limestone
Carnallitic Marl (20 m)
   (2) Middle Evaporites: (130 m)
      Upper rock salt
      Potash rock
      Lower rock salt
      Rock salt and anhydrite
      Anhydrite
Upper Magnesian Limestone (65 m)
   (1) Lower Evaporites: (350 m)
      Upper rock salt with anhydrite
      Upper anhydrite
      Lower rock salt with anhydrite
      Lower anhydrite
Lower Magnesian Limestone (125 m)

Figure 17.8. Make your own diagram based on the facts set out here. Choose a separate colour for each of the stages 1–5 on the left, and draw the rock thicknesses of the borehole succession to scale. Then write down what you have learnt about the drying up of the Zechstein Sea.

conclusions from studies of rock magnetism, suggests that Britain was farther south in the belt of the Trade winds. This evidence is in line with that suggesting a tropical climate for the formation of the Coal Measures, and the presence of glacial deposits of Permo-Carboniferous age in South Africa. The theory of Continental Drift suggests that Britain was a little north of the Equator, and South Africa at the South Pole. Some places, such as the south-east Devon coast, have thick pebble beds resulting from increased erosion: these

Budleigh Salterton Pebble Beds are formed of debris from northern France, and it is supposed that a river brought the spread of gravels northwards.

   The overlying Keuper deposits reflect the fact that the landscape had been worn right down by the later part of the Triassic period. Fine sands are overlain by the red, sticky mudstones which provide the basis for the heavy soils of the Midlands, Cheshire Plain and Trent Valley. The 650 m of so-called 'marls' are repeated alternations of clay, dolomite and evaporite horizons due to slight fluctuations in the humidity of the climate. Fossil scorpions and drought-loving vegetation have been extracted from these layers. The landscape of the period must have been low-lying but hummocky, and the hollows were filled by evaporating lakes. The rock salt horizons in Cheshire are mined for use in the Merseyside chemical industry, and subsidence of the surface is common where the deposit has been extracted as brine.

The Rhaetic Transition

A thin but persistent group of rocks separates the continental Triassic rocks from the marine Jurassic. They are seldom more than 20 m thick but occur throughout the outcrop from Dorset to Yorkshire and record the transition from one environment to the other.

The first advance of the seas merely flooded the lagoons and brackish lakes that existed at the end of the Triassic, and many of the reptiles, amphibians and fish living in them were killed off to form a very thin 2–5 cm bone bed. This **Rhaetic episode** is paralleled by deposits thousands of metres thick in the Alpine area, but in Britain there are usually less than 16 m of varied shales and limestones, including the Cotham 'marble' with its darker, tree-like markings probably caused by the action of algae. Rhaetic rocks are best exposed on the shores of the Severn estuary.

Life in the Upper Palaeozoic

The Devonian and Carboniferous periods saw far-reaching changes in the pattern of life on the Earth. The vertebrate group became prominent for the first time, and life moved out on to the land with increasing success.

The Devonian period saw the real beginning of both these processes. British rocks of this age contain few fossils, but very important ones. Fishes of all types—jawless ostracoderms, placoderms with primitive jaws, sharks and early bony fishes—inhabited the seas and freshwater lakes. The first amphibious creatures emerged, and the earliest insects joined them on the dry land. Both were preceded, however, by the spread of land plants. One of the Old Red Sandstone outcrops at Rhynie in north-east Scotland is a peat deposit of the time, which was impregnated by silica-rich waters from a near-by hot spring, and has preserved some of the most primitive types of land plant (Figure 9.28). By the end of the Devonian there were dense forests. In the seas coral reefs and spiriferid brachiopods were common, and the goniatites—ancestors of the ammonites—travelled widely.

The change to extensive, shallow, lime-rich waters in the Lower Carboniferous led to a flourishing group of corals, productid brachiopods, crinoids and bryozoans. At times they were so abundant that they formed a large proportion of the bioclastic limestones which have resulted. The main difficulty here is that none of them are really adequate as zone fossils, and the Lower Carboniferous has to be zoned by a combination of them all—ie a fossil assemblage named after a distinctive coral or brachiopod it contains. The swamp forests of the Upper Carboniferous demonstrate how complete had been the colonisation of wet-soil conditions by plants, many of which grew to gigantic sizes (over 50 m high). Large insects, including dragonfly-like varieties with a wing-span of 75 cm, as well as spiders, scorpions and millipedes also inhabited these areas, and the lakes contained numerous lamellibranchs. Fishes became more and more common, the amphibians reached the height of their development and the first fossil reptiles are also found in these rocks.

The Upper Carboniferous Coal Measure rocks were another headache for the zoning palaeontologist until recently, when it has been found—rather unexpectedly—that the lamellibranchs which lived in the deltaic swamp pools, as well as plant remains and microscopic spores, can be used with some degree of reliability, especially in conjunction with each other. In addition the occasional marine bands, which are more frequent in the lower horizons, are found over wide areas, linking up several isolated basins of deposition, and they can be used as 'index horizons'.

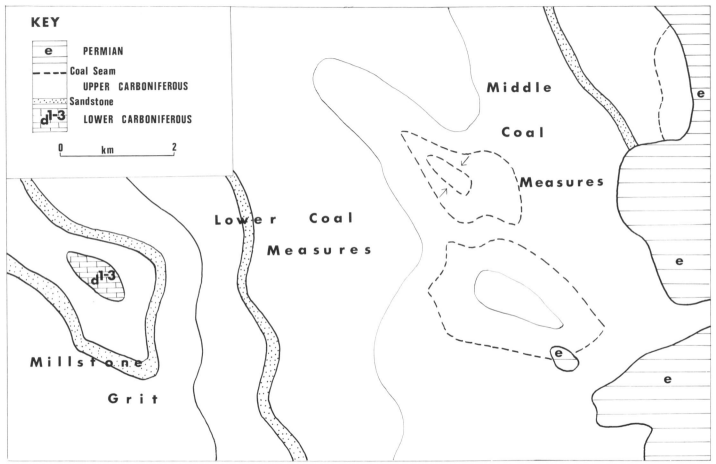

KEY

e	PERMIAN
— — — Coal Seam	
	UPPER CARBONIFEROUS
Sandstone	
d1-3	LOWER CARBONIFEROUS

0 ————— km ————— 2

Middle

Coal

Measures

e

Lower Coal

Measures

e

d1-3

e

Millstone

Grit

e

e

Figure 17.9 Map exercise. Summarise the evidence for (*a*) folding and (*b*) an unconformity, shown by the patterns of rock outcrops in this area. (This sketch-map is based on the Geological Survey 1 : 63 360 Chesterfield Sheet. *Crown Copyright Reserved.*)

The Palaeozoic–Mesozoic Transition

The British desert deposits of Permian and Triassic age provide little evidence for the development of life during these periods, but elsewhere in the world great changes took place between these periods, especially in the seas, and the major boundary between the Palaeozoic and Mesozoic is placed here.

At the end of the Permian many of the animals typical of the Palaeozoic seas became extinct. The last of the trilobites, the rugose corals, the goniatites and many primitive varieties of other groups ceased to exist. The brachiopods and crinoids in particular were greatly affected, and the way was opened for the more intensive development of the better-equipped invertebrate groups, the molluscs and echinoids, which largely replaced many of the Palaeozoic animals. The Triassic began a new era in the story of life on Earth, but there is no record of marine fossils of this age in Britain.

On land there was less change. Plants of the Upper Palaeozoic type gave way slowly to more advanced forms, and the development of the vertebrate animals continued. During these two periods the advantages which the reptiles possessed gave them the mastery of land conditions and of the less adaptable amphibians. By the Triassic the reptiles had not only conquered the land environment but some had gone back to living wholly in the water. Others developed features which resemble the early mammals, such as a single jawbone, different types of teeth specialised for tearing meat and grinding plant matter in the same palate, and a separate nasal passage: it is thought that the first mammals developed at this stage.

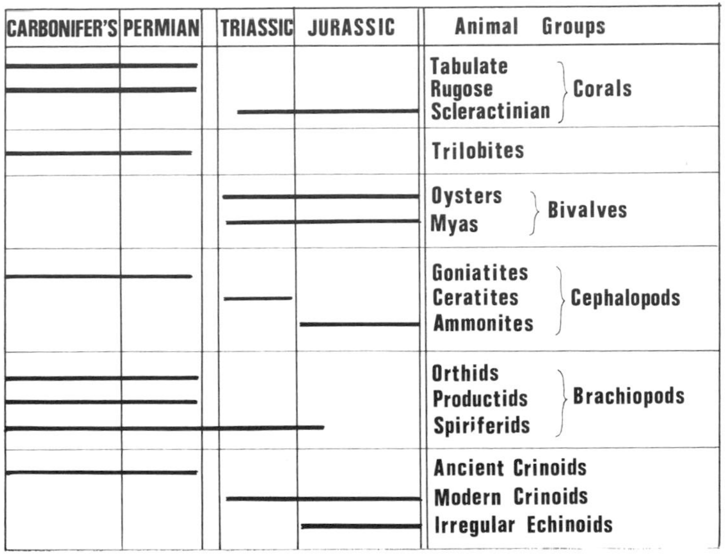

CARBONIFER'S	PERMIAN	TRIASSIC	JURASSIC	Animal Groups
				Tabulate / Rugose / Scleractinian — Corals
				Trilobites
				Oysters / Myas — Bivalves
				Goniatites / Ceratites / Ammonites — Cephalopods
				Orthids / Productids / Spiriferids — Brachiopods
				Ancient Crinoids / Modern Crinoids / Irregular Echinoids

Figure 17.10 The Palaeozoic–Mesozoic division. A number of animal groups became extinct at the end of the Palaeozoic and were replaced by new groups in the Mesozoic, as shown by this diagram. There are other groups, however, which were scarcely affected—eg the vertebrates and the plants.

253

18 The Evolution of the British Isles: the Oldest Rocks

The highest parts of the British Isles—Wales, the Lake District, south-eastern Ireland, the Southern Uplands and the Scottish Highlands—are composed of ancient rocks. Most of them are highly folded and some have been metamorphosed. For these reasons they are amongst the most difficult to study, and many problems of interpretation still exist. It is possible, however, to discern the main outlines, although these grow fainter as the story is traced farther back in time. There is a very interesting comparison here between the oldest rocks of all, the unfossiliferous Precambrian, and the first rocks with an adequate group of fossils, the Lower Palaeozoic group. The former are at their most important in northern Scotland, whilst the latter dominate the other mountainous areas.

Precambrian Time: the Dawn of Geological History
In spite of the fact that these are the Earth's oldest surface rocks, they are frequently found without any rocks of later age on top of them, and 20 per cent of the world's continental areas have Precambrian rock outcrops at the surface. They form the great **shield areas** of low relief: the most extensive is in north-eastern Canada, but all the continents have a stable foundation of this type, which has been more or less covered by more recent rocks. Most Precambrian rocks are metamorphosed schists and gneisses together with many granites, and they often contain rich mineral deposits (iron, gold, uranium, copper, nickel). Recent studies of the Precambrian rocks have revealed that they are not the remains of the first crust of the Earth, because they contain evidence of several phases of mountain-building and volcanic action. The Principle of Uniformitarianism can be taken back to the earliest period of Earth history.

Fossils are great rarities in Precambrian rocks, so that we have a very poor record of what life was like on Earth until 600 million years ago. Algal-like structures have been recognised in these rocks for some years, but in 1959 a group of soft-bodied animals was found in late Precambrian Australian rocks. There was life at this time, but the evidence has either been destroyed or the likelihood of preservation was decreased by the fact that the animals did not secrete hard shells at the time.

The absence of fossils, and the greatly altered nature of the rocks, has meant that they have been largely ignored by geologists until very recently, when great efforts have been expended in their study, using techniques like radioactive dating to help unravel the answers to the great problems of interpretation.

The British Precambrian Rocks
As in many other parts of the world, the British Precambrian includes both metamorphosed and unmetamorphosed rocks.

Scotland is made up of a series of faulted blocks, as Figure 18.1 shows, and most of the country north of the Midland Valley is dominated by Precambrian rocks. There are three distinct assemblages.

(1) The **Lewisian group** is found only in the extreme north-western Highlands and in the Hebrides. It is composed of various types of gneiss, the commonest being similar to granite in composition. These gneisses have been intensely folded and faulted and have suffered further igneous intrusion. The oldest British rocks are in this group and have been dated as 2700 million years old by methods based on the radioactive minerals they contain. At least two periods of mountain-building have been recognised, since some of the intensely folded rocks are cut by unaltered igneous dykes, which elsewhere have been metamorphosed with the surrounding rocks. The first of these is the older (why?). They have been dated by radioactive methods at 2460 million years, and 1500–1600 million years respectively. Look at Figure 11.5 again.

(2) There are rocks forming a complete contrast in the same area:

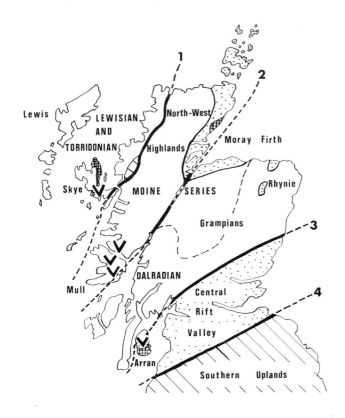

Figure 18.1 The geology of Scotland. A map showing the outcrops of the main groups of rocks in Scotland. The unshaded areas have Precambrian rocks at the surface; oblique shading signifies Lower Palaeozoic rocks; the stippled areas have Old Red Sandstone and Carboniferous rocks; the small areas of heavy cross shading have Mesozoic rocks; the 'V' symbol marks the centres of Eocene volcanic activity. 1–4 are the major faults and thrusts: 1 is the Moine Thrust; 2 the Great Glen fault; 3 the Highland Boundary fault; 4 the Southern Uplands Boundary fault.

unmetamorphosed red arkose sandstones with some conglomerates and shales are over 3000 m thick in places. These **Torridonian** rocks lie on the Lewisian with a marked unconformity (Figure 18.2) and must have been formed after a long period of uplift and erosion, for the rocks beneath the unconformity are the deepest roots of the mountain ranges that once stood there. The Torridonian sandstones contain fragments of feldspar minerals, which are easily weathered in humid conditions, and wind-faceted pebbles and were probably formed on the extensive flood-plains of rivers draining a land mass with restricted and seasonal rainfall (semi-arid conditions). Much of the debris came from the north-west—an area now covered by the Atlantic Ocean.

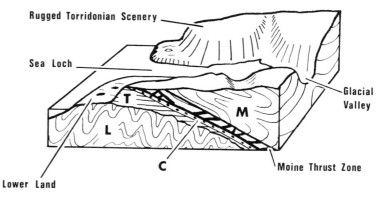

Figure 18.2 North-west Scotland. Try and recognise two unconformities. L = Lewisian; T = Torridonian; C = Cambrian; M = Moine Series.

(3) The majority of the North-west Highlands are formed by the **Moine Series** of rocks, which extend southwards into the Grampians. Like the Lewisian they have been strongly metamorphosed, and are even more monotonous in rock-type. They are almost entirely of

255

The extreme north-west of Scotland. The flat, monotonous surface of Lewisian gneiss in the foreground is broken by the hills of Torridonian sandstone rising on the skyline. (*Crown Copyright*)

sedimentary origin; wide areas are characterised by flaggy, silica-rich granulites, and variations are rare. The whole series has been thrust westwards over the Lewisian and Torridonian rocks (Figure 18.2). In one or two places the Moine rocks rest unconformably on the Lewisian, but radioactive dating is complicated by the fact that the rocks were last involved in folding and metamorphism in the Caledonian orogeny and so record that date (ie late Ordovician, *c* 430 million years ago, in this area). Some areas of Moine rocks near the western margin of their outcrop seem to have been sheltered from

these movements, and intrusive igneous rocks cutting across them have a date of 740 million years. The greater part of the Moine rocks may have been formed about 1000 million years ago. It is now accepted that these rocks have a similar age to the Torridonian, but were formed in different conditions and have undergone more intense deformation.

These Scottish rocks illustrate many of the stratigraphic problems posed by very old rocks. What were the original rocks before metamorphism took place? How can we tell whether such rocks in two widely separated areas are of the same age? What do the radioactive dates tell us?

Rocks of the same age in England and Wales add little to the picture presented by the Scottish outcrops: they are so small in extent and widely scattered. The highly metamorphosed groups are paralleled in Anglesey and the heart of the Malvern Hills, but elsewhere there is a variety of sediments as well as volcanic lavas and ashes. The most important outcrops are in Anglesey and the Longmynd massif of Shropshire, and there are smaller occurrences at Charnwood Forest (Leicestershire), in west Pembrokeshire and in the central Pennines.

The Lower Palaeozoic Orogenic Cycle

Whilst the Precambrian rocks record the existence of even earlier orogenic cycles, their evidence is partial and extremely complex: in many cases the rocks have been subjected to later phases of alteration or have been involved in later mountain formation. During the Lower Palaeozoic, for instance, the centre of rock formation and orogenic activity lay across the British Isles, as Figure 18.3 shows, and many of the Precambrian rocks were caught up in the events taking place.

The pattern of events in the Lower Palaeozoic, as recorded in the rock outcrops of the British Isles, is not quite so simple as that we studied in connection with the Upper Palaeozoic geosyncline in Chapter 17. Several features are particularly important in connection with the Lower Palaeozoic evidence.

(1) There seem to have been **several troughs** within the main area of subsidence, which extended between the far north-west of Scotland and the English Midlands: Figure 18.4 shows the positions of these troughs, which were mostly orientated from north-east to south-west. The areas between either have no sediments of Lower Palaeozoic age preserved today or can be shown to have been land which supplied debris to the troughs during that time. Anglesey, for instance, was a source of sediment for the Welsh trough.

(2) **The troughs developed at different times.** The most northerly began earliest with the formation of the Dalradian group of rocks, probably starting in the late Precambrian and continuing into the Cambrian.

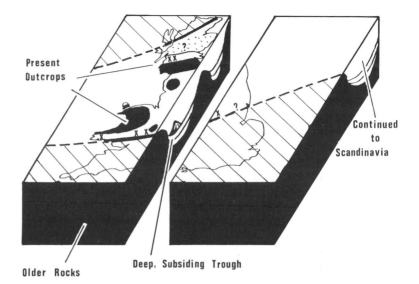

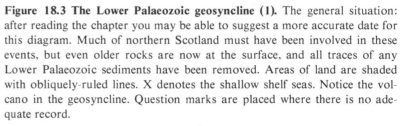

Figure 18.3 The Lower Palaeozoic geosyncline (1). The general situation: after reading the chapter you may be able to suggest a more accurate date for this diagram. Much of northern Scotland must have been involved in these events, but even older rocks are now at the surface, and all traces of any Lower Palaeozoic sediments have been removed. Areas of land are shaded with obliquely-ruled lines. X denotes the shallow shelf seas. Notice the volcano in the geosyncline. Question marks are placed where there is no adequate record.

Uplift and metamorphism with intrusion took place in this trough from the Middle Ordovician through to the Devonian. Farther south sedimentation began only in the Cambrian and continued into the late Silurian before uplift caused a marked break. There is also a progression of lessening intensity of earth movements from north to south.

(3) **Each trough had its own distinctive history of sedimentation.** Some became filled in at certain periods and land appeared. This was

257

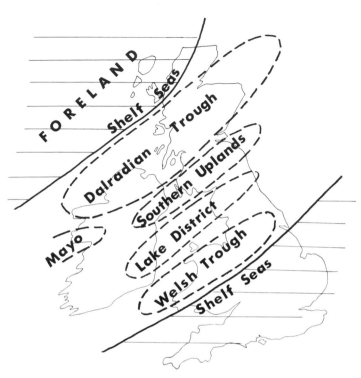

Figure 18.4 The Lower Palaeozoic geosyncline (2). The main troughs of sedimentation are shown. This area can be linked north-eastwards to Scandinavia and westwards to North America. In each area there are similar rocks of this age (see Figure 14.10).

particularly common during the volcanic outpourings of late Ordovician age in the Snowdonia region of north Wales, and of Middle Ordovician age in the Lake District: submarine pillow lavas gave way to subaerial tuffs and ignimbrites.

(4) **Different groups of fossils** are found in the two main types of sedimentary facies. The deep-water sediments of Ordovician and Silurian age are characterised by graptolites, whilst the shelly sediments of the margins contain trilobites, brachiopods and tabulate corals. Comparable facies occur in the Upper Palaeozoic geosynclinal sequence, but the fossils are quite different: goniatites replace the graptolites as swimming forms found in deep sea sediments, and other varieties of coral and brachiopods are common in the shallow water sediments.

The Dalradian and Cambrian Rocks

These two groups overlap in time and are the basal members of the piles of sediment involved in the Lower Palaeozoic geosyncline.

The **Dalradian Series** overlie the Moine Series in the southern and eastern Grampians and extend into northern Ireland. They are a series of sediments which have been metamorphosed, but the original structures are often present. The rocks include quartzites, possibly tillites, and algal limestones formed in shallow waters, deep-water dark-coloured shales and thick greywackes and volcanic lavas and ashes. One of the uppermost layers contains Cambrian trilobites, and this group of rocks seems to span the transition from Precambrian to Lower Palaeozoic times. The radioactive dates are similar to those of the Moine rocks, since both were involved in the same orogeny.

It seems that the Moine Series was a group of much older rocks, already folded, uplifted and eroded, which became involved as basement to the Dalradian Series: renewed burial led to renewed metamorphism.

The few outcrops of **Cambrian rocks** in Britain show us the beginning of the story elsewhere, as the geosynclinal seas which already covered Scotland extended over the worn-down remnants of the Precambrian landscape. The rocks formed in these seas rested on the folded and eroded Precambrian basement with a marked unconformity. There are pebbly conglomerates nearly everywhere at the base of the series of rocks, often succeeded by hard orthoquartzite sandstones and then by fine-grained shales. In Pembrokeshire, near St

Davids, the conglomerate pebbles are kidney-coloured quartz, and this layer is covered by green, red and mauve sandstones, all unfossiliferous. The overlying shales contain many trilobites. In this area, and elsewhere throughout the country, the later rocks cover an increasing area and become finer-grained—evidence that the sea was transgressing across the land, as Figure 18.5 shows.

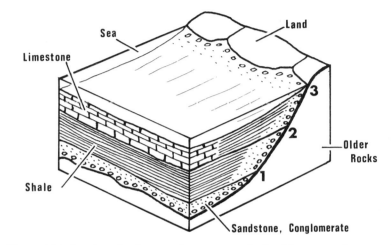

Figure 18.5 Overlap. As the sea advances over the land it covers a larger and larger area—with shorelines at 1, 2 and 3—and so later deposits overlap earlier ones. Notice how the succession of rock-types will vary, and the differences within a layer of rock formed at one period of time.

We must go to north-west Scotland to examine the northernmost group of Cambrian rocks in Britain. There conglomerate and ortho-quartzite rest unconformably on the Torridonian, from which they are separated by the usual unconformity (Figure 18.2). There must have been uplift and erosion intervening between the formation of the Torridonian and Cambrian rocks. The overlying Durness Limestone, which is a dolomite, contains trilobites which show the rock to be partly Cambrian and partly Ordovician in age, but they are different species from those found in the shallow-water deposits of south Wales. They resemble more closely trilobites which are typical of the North American Cambrian rocks. We have already seen that there was a continent to the north-west supplying debris for the Torridonian deposits, and North America may well have been closer to us at that time if the theory of continental drift is correct.

The rocks of north Wales are mainly fine-grained, and have been altered by subsequent folding into slates which are quarried for roofing materials. The few outcrops of Cambrian rocks in the English Midlands are mostly shallow-water sands, but with thicker shales towards the north. There is thus a pattern of shallow-water deposits to the north and south, with deeper-water and finer sediments forming in between. Animals living in the shallow waters could move along the margins of the land in north-east to south-west directions but could not cross the intervening deeper stretches of water. The northern and southern margins therefore have different species of fossils in the rocks.

The Ordovician and Silurian Rocks
The Cambrian rocks occupy only a very small area at the surface and present a fragmentary picture which we could not reconstruct with any confidence unless we had a fuller record of the Ordovician and Silurian rocks whose outcrops are more widespread. It is convenient to consider the Ordovician and Silurian together, since they have many common features, and were originally placed in the same major division of geological time.

The **Ordovician period** saw many important events in the development of the geosynclinal trough. It seems that the seas were at their deepest in this period, for in areas like the Southern Uplands and Wales the Ordovician beds are black shales, indicative of deep, stagnant waters some distance from the land. These rocks contain many fossils of planktonic graptolites, which enable accurate correlations to be made between the regions. The margins of the troughs can also be plotted, because there are areas with a quite different group of

Snowdon summit (centre): a syncline of Ordovician volcanic rocks eaten into by more recent glacial features. (*Aerofilms*)

sediments and fossils. On the Ayrshire coast, where the northern edge of the Southern Uplands meets the sea, and again on the Welsh borders, there are sandy and calcareous sediments containing a wealth of trilobite, brachiopod and mollusc fossils—all of which lived in shallow seas. There were at least two types of sedimentation at this time: one in the deep troughs and the other in the surrounding shelf seas. The rocks and the fossils they contain represent two distinctive facies, and since they overlap in some places it has been possible to correlate the 'graptolitic' and 'shelly' groups of rocks.

The development of the geosynclinal trough appears to have included south-easterly movement of the main area of subsidence. Already during the Ordovician period folding and metamorphism were taking place in the Scottish part of the geosyncline, and mountain-building may have begun, resulting in uplift and emergence of the newly formed rocks. This led to erosion and the formation of local unconformities in many places. There was also a lot of volcanic activity associated with such movements. Britain must have been in a zone of fire rather like that which girdles the Pacific Ocean today

(Figure 10.1), with many volcanoes erupting lava and ash. Most of this activity took place on the floor of the geosyncline, giving rise to thick masses of pillow lava and solidified tuff interbedded with the ordinary sediments. The rock types vary from the rhyolites of Snowdonia to the andesites of the Lake District and the more basic spilites. The resultant rocks are very resistant and stand out in the rugged scenery of the Snowdon area in north Wales (where the lavas and ashes are over $1\frac{1}{2}$ km thick) and of the central part of the Lake District (where they are nearly 3 km thick).

Towards the end of the Ordovician period some of these areas record a gradual infilling of the trough with coarser deposits due to more rapid erosion of the surrounding lands including the newly formed fold mountains.

The **Silurian period** led up to the climax of the geosynclinal series of events. Although the volcanic activity almost died out, the familiar pattern of deep geosynclinal trough and shallow bordering shelf seas was continued, at least in the early stages. An increasing amount of earth movement began to affect the areas of deposition as the period advanced, leading to increasingly rapid erosion of the marginal continents and rapid infilling of the troughs with coarse conglomerates, greywackes and flagstones, outpacing any deepening that was still taking place. This began earliest in the Southern Uplands and the Lake District, and later affected Wales, where thousands of metres of these sediments were rapidly piled up.

The movements that took place in the Silurian led to an extension of the shallower shelf seas eastwards over the English Midlands, and a series of sandstones, limestones and shales were deposited on this sea floor in alternating clear and muddy conditions. The Wenlock Limestone, which has been tilted and eroded to form the Wenlock Edge escarpment, contains a wealth of 'shelly' fossils—brachiopods, trilobites, corals and crinoids. This is in the middle of the Upper Silurian sequence, but towards the top there is evidence that the seas were being filled in here as well. The rocks are current-bedded, and the fossils are reduced to a very few varieties which could put up with the difficult conditions in a drying, shallow and possibly highly saline sea.

At the very top of the Silurian rocks in the English Midlands there is a thin layer known as the Ludlow Bone Bed, packed with the remains of the most primitive fish-like vertebrates. Was this formed by the wholesale death of these animals in a great catastrophe, or did it accumulate slowly over the years whilst sediment was swept away by strong currents? The answer is not clear. Compare this bone-bed horizon with that found in the Rhaetic (Chapter 17). How different would the fossils have been in the two deposits?

Life in the Lower Palaeozoic Seas

The developing geosyncline saw a remarkable growth of life on the Earth. All the animals and plants of which we have a record lived in the sea, and they were dominated by primitive representatives of the invertebrate groups. There was a surprising variety of animals, and all the major groups (ie phyla) have fossil records extending back to these rocks. The Ordovician period in particular was one of the most important for the emergence of new groups of animals.

The most important Cambrian zone fossils are the trilobites, which included some of the largest varieties, and the primitive, inarticulated brachiopods. Many other groups, such as the echinoderms and molluscs, are also represented by fossils of their most primitive known species. Sponges, coelenterates, algae and even worms have been found in Cambrian rocks.

The Ordovician saw many changes and the first appearance of some important fossil groups. Trilobites are found in even greater numbers and have a great variety of shapes adapted to a wider set of conditions. Brachiopods, too, developed many new forms with properly articulated shells. The graptolites, seemingly small and insignificant, were the most important of the newcomers and have become the main zone fossils for the rocks of this and the following Silurian period: their shapes change for every few metres of rock. In some of the lime-rich shelf seas the first of the true corals, as well as echinoids, bryozoans, molluscs and crinoids flourished. Long, straight cephalopods, up to 4 m long, and the eurypterid 'sea scorpions' were the largest animals.

Two escarpments formed of Silurian rocks in Shropshire. To the left is Wenlock Edge (Wenlock Limestone), and to the right the feature formed in Aymestry Limestone. (*Aerofilms*)

The Silurian did not produce so many changes but continued the development of a rich and varied life in the seas. The Wenlock reefs illustrate the diversity present in the British shelf seas. In addition the upper Silurian Ludlow Bone Bed contains the earliest British remains of primitive vertebrates—the fish-like ostracoderms. These may have developed in freshwater conditions and therefore would have left a poor fossil record of the early stages of their evolution. Earlier fragments of such fossils have been found in Ordovician rocks in Wyoming (USA).

The Caledonian mountain-building caused great changes in the geography of Britain and hence in the habitats available to plants and animals. Many changes were also caused by the increase of the vertebrates and the possibilities of life on dry land. The Silurian rocks saw the last of the graptolites, of the trilobites as a group of major significance, and of many of the more primitive members of major fossil groups.

The Caledonian Orogeny
When the formation of new rocks came to an end, the upheavals which

had been affecting sedimentation during and after the Ordovician finally resulted in a vast series of mountain ranges. Great Britain lay across the path of violent movements as the episode of mountain-building known as the **Caledonian Orogeny** reached its climax. Great chains of mountains running from south-west to north-east across Britain and Scandinavia were produced from the rocks which had been accumulating in the geosyncline: where there had been a deep trough in the sea, developing over a period of 200 million years, there were now high ranges of folded mountains and almost the whole area became land. Up to 11 km of sediment had been deposited in the geosyncline during that time, constituting a vast mass of relatively light crustal material. Compressional forces within the Earth's crust plus isostatic compensation now led to the crumpling of these rocks and their uplift.

The effects varied from one area to another both in time and in intensity. In northern Scotland there was considerable alteration, probably during the Ordovician, of the rocks we call the Moine and Dalradian Series, and they were intruded by masses of granite and gabbro during and after the Silurian, so that any results of earlier earth movements affecting these rocks were completely masked. The intense pressures caused tight isoclinal and large-scale recumbent folds to be formed, and the whole mass was thrust against the resistant ancient rocks of the north-west (Figure 18.2). It is particularly difficult to puzzle out the structures produced by these earth movements, but it seems that the rocks were more plastic and easily deformed at first, when larger folds covering many kilometres of country and small-scale corrugations were formed in the rocks. A later phase of activity affected the more brittle rocks by faulting and thrusting. The metamorphism of these rocks took place at this time and locally included migmatisation. This is why these rocks give a radioactive date coinciding with the Caledonian movements. Today we see the once deeply buried 'roots' of the Caledonian Mountains, which have been eroded and uplifted again and again during the geological history of the area. Remember this fact as you study the later periods in the development of our islands.

A great rift valley opened up across central Scotland, which was to form an important feature through the ages. The fine shale deposits of southern Scotland were folded in closely packed isoclinal folds, but there was little metamorphism apart from the immediate neighbourhood of some of the granite masses intruded at this time of upheaval. The graptolite fossils contained by these rocks are still well preserved. Perhaps erosion has not reached the deepest mountain 'roots' here, as it has done farther north.

To the south, in the Lake District and Welsh areas of the geosyncline, the effects now revealed at the surface were less dramatic. Folding was on a broader scale, partly because of the greater proportion of more resistant volcanic and greywacke rocks and partly because the pressures were not quite so great. Metamorphism was on a less intense scale as well though it affected wide areas. The more intense changes are confined to the aureoles of intrusive granites and gabbros in the Lake District and to local areas where weaker shales were compressed to form slates, as in the Llanberis and Blaenau Ffestiniog areas of Wales.

Not only were the most intense effects of the orogeny experienced in the far north, but they were received there first. We cannot be sure of the age of formation of many of the rocks in the Scottish Highlands, but we know that sedimentation continued at least into the Ordovician (Durness Limestone) and that it finished earlier in the Southern Uplands than farther south. It must be remembered that the movements were not limited to a narrow period of time, but that only the greatest degree of uplift took place between the formation of the last Silurian and earliest Devonian rocks: other movements began in the Ordovician and carried on into the Devonian.

The Old Red Sandstone: Post-orogenic Deposits

The Caledonian upheavals replaced the seas by mountain chains and inland basins of deposition (Figure 17.1). Three areas of land sediments can be distinguished: the south-eastern Welsh borders, where fluvial deposits gave way southwards to the Devonian seas; the central Scottish rift valley; and the lacustrine basin of north-eastern Scotland.

The **Welsh rocks** are a mixture of red and green sandstones, with shales and conglomerates. The typical, repeated sediments are those of a fluviatile plain (Figure 18.6.1) with channel lag, point bar and overbank flood deposits (see Figure 5.15). The climate must have been sufficiently rainy to give rise to at least seasonal streams, shifting their courses across the coastal plain.

1

Fine Silt

Fine Sand

Cross-bedded Sand

Pebbles

2

Cross-bedded Sand

Black Flags

Limestone

Mudstone

3

Sands with Pale Muds

Bituminous Shales

Mudstone

Figure 18.6 Rhythmic sedimentation in the Old Red Sandstone. Relate these sequences to the manner of their formation, as suggested in the text.

In **Scotland**, the **central rift valley** was the site of coarser wedges of sediment filling the depression from the erosion of the mountains to north and south. Great alluvial fans built up along the margins, similar to those occurring in semi-arid mountain regions today. The deposits here locally reach over 6000 m in thickness and are interbedded with lavas, mainly andesites and basalts plus tuffs and agglomerates. The fossils in the Welsh and central Scottish rocks are similar: early plant and vertebrate remains. Other areas, marginal to the central rift, also experienced igneous activity.

At Ben Nevis and Glencoe these lavas have dropped down in cauldron subsidence and are surrounded by ring dykes: the process brought them into close proximity with large granite masses, which were being intruded at the time. Thus Britain's highest mountain summit is actually made of a downfaulted mass of lava. Some of these granites are associated with swarms of south-easterly trending dykes. The Cheviot, farther south, was an active volcano (Figure 18.7).

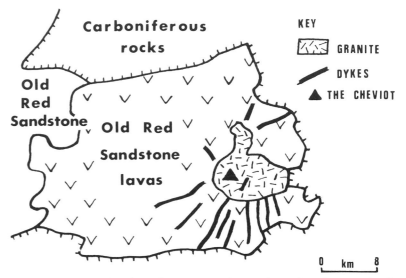

Figure 18.7 The Cheviots. Can you work out the order of events in the formation of this complex of igneous rocks? If not look back at the section 'Igneous Rocks and Time' at the end of Chapter 10.

In the **far north of Scotland** another basin of Old Red Sandstone deposition has been preserved round the Moray Firth, in Caithness and in the Orkney and Shetland Islands. Thick deposits of grey, flaggy sandstone form the striking cliffs of Caithness and these thin layers of rock contain groups that are repeated again and again (Figure 18.6). They must have been deposited on a sinking plain, and conditions determined whether limestone, bituminous shale or sandstone was formed. There are scattered outliers of these rocks on the Grampians to the south of the Firth.

THINGS TO DO AND DISCUSS, CHAPTERS 15–18

Some topics for discussion and essay-writing

(a) Write notes on three of the following formations: Wenlock Limestone, Lower Greensand, Torridonian Sandstone, London Clay. In each case, refer to the geological period to which the rock belongs and the conditions in which it was formed, say where it can be seen in Great Britain, and describe its most important characteristics, including typical fossils (S).

(b) Describe in detail a common fossil from either (i) the Chalk or (ii) the Carboniferous Limestone Series. (S).

(c) Describe the geological history of a region in which occurs the following sequence of rocks (A is the oldest and I the youngest):

I. A series of NW–SE dykes cuts all the strata of the region.

H. Occasional outliers of pure white limestone, containing echinoids and flints, rest unconformably on G.

G. A series of massive limestones, containing ammonites, brachiopods and some corals, overlies the earlier rocks, unconformably.

F. A series of east–west dykes cuts all strata from A to E, inclusive.

E. Sandstones and shales, with coal seams.

D. Limestones, containing corals, passing upwards into alternating grits and shales, with goniatites.

C. Muddy limestones, rich in brachiopods and trilobites, resting unconformably on both A and B.

B. Unfossiliferous red arkose sandstone, overlying A unconformably.

A. Schists and gneisses. (O.)

(d) Describe, using examples from British stratigraphy, the main features in rocks which indicate a period of marine transgression (O).

(e) Describe the principles upon which the stratigraphical column has been drawn up (O).

(f) Describe briefly the conditions of formation of (i) Lower Palaeozoic, and

(ii) Mesozoic rocks, in Great Britain, commenting on their contrasting fossil content (O).

(g) Describe the environment in which the deposits of the New Red Sandstone (Permo-Triassic) were formed (N).

(h) Indicate what is meant by the following, selecting specific examples from the stratigraphic record:

(i) arid conditions, (ii) deltaic deposits, (iii) a change in facies, (iv) a marine transgression (AEB).

(i) What evidence is there to suggest that sedimentation in Cornwall and Devon differed from that in other British areas during Devonian times? (L).

(j) Explain how fossils are used to correlate rocks of the same age found in adjacent regions. What fossils have been used for this purpose in Palaeozoic and Mesozoic rocks? (C).

(k) Discuss the use of fossils in stratigraphy (W).

Part Five

Practical Geology

19 Field Geology

Every geologist must be a field geologist, since all the material for his scientific observations is to be found in the rocks of the coast and countryside around him. Those who use their geological knowledge to earn a living make this the basis of all their decisions—they have to go to the place in question before drawing any conclusions. Much of the analysis of results and identification of specimens can be done only with the facilities of a laboratory or library at hand. But the basic work of the geologist must be done in the field, and there is plenty of scope for the amateur geologist to investigate the rocks in his own locality, provided that it is done in the correct way.

Equipment Needed

A geologist's field equipment can be very simple and cheap. He will need some old but warm and rainproof clothing, a notebook and mapcase, a hand lens and implements with which he can break open the rocks to make investigation easier. It is important to examine unweathered surfaces of a rock as well as those exposed in the outcrop, and to prise out the fossils. The hardest rocks will only yield to heavy hammers weighing 1 kilogram or more, but many soft rocks will need lighter and more delicate treatment. Cold chisels are useful for extricating fossil specimens from hard rocks; a broad-bladed penknife from soft. A clinometer is used to measure the rock dips, and a tape-measure for thicknesses. Tins and boxes are essential for preserving pieces of fragile rock or fossil, many of which will need to be wrapped in paper or even cotton-wool for the journey home. All this equipment can be carried in a light rucksack.

The Needles and Alum Bay, Isle of Wight. (*Aerofilms*)

In addition the geologist will need storage space for the specimens he has collected and labelled, and books of reference to assist him in the identification of the rocks, fossils and minerals. A list of these is given at the end of the book.

Where to Look

The solid rocks are often masked by glacial deposits, soil, landslips, peat, vegetation and the works of man. Inland exposures are largely confined to natural scars, which may be abundant in upland areas, quarries, river banks and fresh road and railway cuttings. Permission must be obtained to examine quarries and railway cuttings. Making a map in areas of complex structure with few exposures is very difficult even if the geologist has records of bores and mine workings to aid him. The line of outcrop of a hard rock can, however, often be traced by means of a relief feature such as a ridge or escarpment.

The best areas for the study of the solid rocks are the cliffed coasts, where there are excellent sections through whole series of rocks, especially where they are steeply inclined. One must always keep away from dangerous crumbling cliffs, and it is best to examine the rocks of the foreshore. Within a space of 3 km on the east coast of the Isle of Wight one can, because they are nearly vertical, study the complete succession of Cretaceous and Tertiary rocks which make up the island and span nearly 100 million years in time (Figure 19.1).

It is best to begin by studying an area of more recent rocks in areas like the Isle of Wight, Dorset or North Yorkshire coasts. Such rocks will be easier to study, will contain more fossils and in general will yield satisfying results more rapidly. Much has been written about these areas, and the guides published by the Geologists' Association give an excellent introduction to what you should find—or a check on what you have discovered.

If you live inland you will be able to examine rock exposures in quarries, etc and will often be able to trace an outcrop across country in the relief. You can also place a greater emphasis on noticing the ways in which geological processes are acting today. Variations in stream-flow and load, evidence for soil creep or for glaciation in the

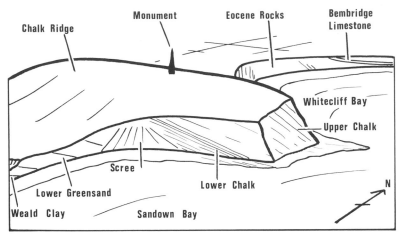

Figure 19.1 The eastern Isle of Wight coast.

Labels on figure: Chalk Ridge, Monument, Eocene Rocks, Bembridge Limestone, Whitecliff Bay, Upper Chalk, Scree, Lower Chalk, Lower Greensand, Weald Clay, Sandown Bay, N

past and the development of features like stream meanders, waterfalls and lake deltas all form profitable lines of study. It is important to make sketches of such features, adding geological notes (see, for instance, Figures 5.4, 6.10, 6.13, 10.6); if you are not an artist a photograph will be just as good.

Observations to Make

Observation is the essence of field-work. Spend time noticing facts about each exposure or section of cliff and record them in your notebook: such a record can take the form of a sketch or photograph as mentioned above, or an orderly list of the details. Throughout this book we have emphasised the importance of doing this and have suggested a number of questions that you can ask yourself. Here are most of them again:

(1) What is the exact position of the exposure? You will need to plot the details on a map later, so make a note or head the page of your notebook with the Ordnance Survey grid reference.

(2) What is the nature of the rocks? Write down a careful description of their colour, texture and the jointing and bedding structures affecting them. Are they sedimentary, igneous or metamorphic?

(3) If they are sediments, what is the dip and the strike of the series of rocks? Does it vary? Are there any unconformities? Are there any folds or faults affecting the rocks, and if so of what type? Is it possible to work out the succession of rocks from oldest to youngest?

(4) If the rocks are igneous, can the margin of the intrusion or lava flow be seen, or is its nearness indicated by smaller crystals? What is the rock-type, and is it a lava-flow, dyke, sill, etc?

(5) If they are metamorphic, is the alteration limited to any zone? See Figure 10.6 for a field sketch of the Cleveland Dyke and see how many of the questions in (3), (4) and (5) are answered.

(6) What fossils do the rocks contain? Do they occur in special bands? The longer you hammer at sedimentary rocks the more you are likely to obtain. They will help you to date the rocks and to understand how they were formed.

(7) What do the landforms of the area reveal about the relative resistance of the rocks to erosion?

(8) What is the evidence for recent geological activity—eg river valley features, glacial landforms, coastal changes?

It might help here to refer you to an exercise given to a group of young geologists and based on the section at the eastern end of the Isle of Wight shown in Figure 19.1. You should be able to answer some of the questions from that diagram. Each student was provided with an outline map (Figure 19.2), and the following list of questions.

(*a*) Make a description of the rocks at each of the places A–I in your notebook.

(*b*) Mark the angle of dip in the rocks where the arrows show the direction on the map.

(*c*) Mark in the boundaries of the rock outcrops on your map and the direction of strike in one place only.

(*d*) Draw diagrams of any fossils you find, and especially of those you cannot take away; note the map position of each discovery.

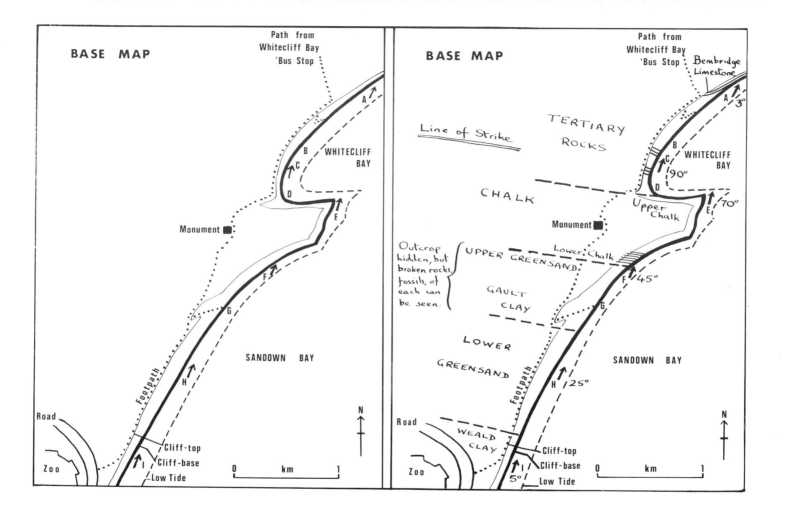

Figure 19.2

Figure 19.3

269

(e) At the end of the day draw a vertical section along the line of the cliffs and work out the structures affecting the rocks. Write a short historical account of the geology of the area.

The map, Figure 19.3, shows the results plotted from investigations on the ground. Your notebook might read as follows:

Point A: northern end of Whitecliff Bay

Rocks dip at 3 degrees to the north

Rocks: massively-bedded (up to 2 m thick), well-cemented, cream or buff coloured; contain many gastropod fossils (freshwater snail type), but most of shells dissolved away leaving casts.
This rock is a freshwater limestone.
It is one of the youngest in the Bay, since it lies on top of the others.
The sea has undermined the rocks here, and there are large broken boulders on the beach.
At the extreme north of the Bay the dip of the rocks brings them down to sea level, and they form a reef extending out to sea.

(Here there would be a drawing of the gastropod and a labelled sketch of the massive bedding and sea-broken blocks.)

Point B: just south of the main cliff path.

Rocks dip almost vertically.

Rocks: softer than those at A, varying from unconsolidated sands to weak clays . . . etc.

Making a Simple Map

You will thus make many observations and collect many specimens as you study the rocks of a particular area, but the most important result will be to make a map of your findings. You have seen what geological maps look like and have learnt to interpret them. Now you should be starting to make your own.

The base-map, on which you will record your information, will normally be available already: the Ordnance Survey 1 : 10 000 are the best, since there is room to plot a lot of information, and you can still see the wider relationships. Remember, however, that the thickness of your sharpened pencil line covers 3 m on the ground. You will often have to use symbols for plotting information on the map: some of those that are used commonly are listed in Figure 19.4.

When the map is ready for use you can record the information gleaned from such questions as we asked above. Detailed sections can be noted at the side of the map or references made to your notebook. The most important features to record at first are the points of contact between the layers of rock, the dip measurements, and the trends of igneous intrusions, faults and folds. Having done this it should be

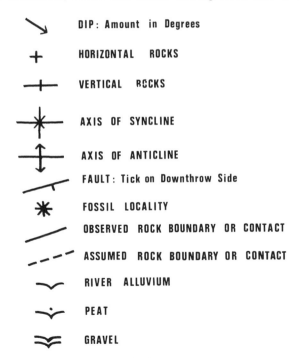

Figure 19.4 Signs used on geological maps.

possible to join up the various boundary outcrops, using the principles learnt in the study of geological maps. Then you should colour the different rock formations, as a means of helping you to interpret the structures in the same way as you have done on the other maps.

TOPICS FOR DISCUSSION AND ESSAY-WRITING

(a) Draw a sketch-map to illustrate the geology of an area which you have studied in the field, and give a stratigraphical table of the rocks which occur there. Discuss the relationship of the topographical character of the area and its geology (W).

(b) Field-work is an essential part of geological studies. Why is this? Describe features of excursions you have participated in which have supplemented your classroom work (S).

(c) Comment upon the main stratigraphical principles illustrated by the rocks of an area you have studied in the field (O).

(d) Under the headings: rock type, structure, and fauna, describe any two exposures that you have visited. Then, from the evidence you provide, attempt to deduce the conditions under which the rocks were formed (AEB).

(e) Describe with the help of sketches a quarry or cliff section you are familiar with, and give an account of all the geological features which can be seen in it (S).

20 Geology in the Laboratory

Whilst the geologist collects his evidence in the field, he is dependent on his laboratory for its equipment and reference books, which will enable him to analyse his observations and specimens in detail, at greater leisure, and without interruptions from the weather. There is not space in this book to do any more than indicate the scope of what is possible, although much has been suggested already at the end of each chapter or section to emphasise the importance of this sphere of geological training. Besides the background reading (a list of Further Reading and reference books follows this chapter) and general discussion, there are several activities which should be part of your geological laboratory work.

1 Examination of Specimens
Specimens of minerals, rocks and fossils can be examined in detail so that you become accustomed to handling them and acquainted with their distinctive features. Investigations relating to studies of this nature are included at the end of Chapters 3, 8, 9, 10 and 11. As you progress you will begin to use instruments like the petrological microscope and can delve into reference books to look up the precise name of a particular specimen. It is important that you learn first to make observations in the most efficient manner.

2 Preparation for, and Follow-up of, Field-work
Time spent in the field is normally at a premium. It is best to make sure that you can use instruments like the clinometer and compass before you begin field-work, and dip readings can be practised on a tilted book. Make sure that you know the Country Code and the Field Studies Code (both obtainable from the Geology Division of Nature Conservancy, Oak Cottage, Brimpton, Reading, Berks) and look up the area you plan to visit on a map, trying to ascertain where the best exposures and view-points are to be found.

When you return from field-work the specimens collected should be checked and labelled neatly from the notes you made about them in the field (ie location, position in exposure, preliminary identification); detailed identifications can now be made. Always write up an account of your field-work observations, together with any associated laboratory work: this account has often proved to be a valuable source of future reference, whereas the field notebook record may be spoilt by the weather on a later outing. If possible thin sections of the rocks you collect can be made, and loose sediments can be analysed in terms of roundness, size and nature of their constituents.

3 Experimental Work
Experimental work is also worth while, although it is not easy to simulate the natural processes in restricted laboratory conditions. If equipment such as a wave tank or stream table is available you may be able to carry out some of the experiments suggested at the end of Chapter 7, and others need only simple apparatus. It is also instructive to learn some of the techniques involved in preparing thin sections of rocks, for instance.

4 Map Analysis
Map analysis is an important technique for the geologist to acquire. Geological maps record the surface distribution of rock outcrops, but this is sufficient to enable the trained student to work out the structures which affect the rocks and a general impression of the main events in the geological history. The subject is covered in greater depth in *Reading Geological Maps*, by M J Bradshaw and E A Jarman, but we mention a few guidelines here.

Outcrop Patterns. The first step involves recognising the distinctive outcrop patterns produced by certain rock structures. Use simple polystyrene models, like those suggested at the end of Chapter 12, to demonstrate these patterns.

(*a*) **Horizontal rocks** have an outcrop which runs parallel to the relief contours (Figure 8.3).

(*b*) **Vertical rocks** will have a straight-line outcrop.

(*c*) **Tilted rocks** (ie between horizontal and vertical) will have outcrops which form V-shapes in valleys and cross the relief contours (Figure 12.4). Draw your own sketch-maps to show the patterns when the rocks are dipping up-valley or down-valley (the latter at gradients which are (i) steeper and (ii) gentler than the valley floor long profile).

(*d*) **Unconformities** result often in outcrops of two different groups of rocks tilted in different directions or at different gradients: the outcrops of the younger group bury that of the older (Figure 8.23 and Figure 12.9), and the line where the two groups meet is known as the overstep.

(*e*) **Folded rocks** affect the outcrops of one group of rocks in which there are two opposing dips (Figure 12.9 and Figure 12.15). The outcrops of individual beds of rock are repeated on either side of the fold axis.

(*f*) **Faulted rocks** are broken along a line, and this is obvious on a geological map because the rocks on either side of the fault line are offset (Figure 12.10). Movements may be vertical or horizontal, and rocks may be repeated, or cut out, on opposite sides of the fault line.

(*g*) **Igneous rocks** include lavas and tuffs, which are bedded with the sedimentary rocks and are considered as part of the general bedded succession. There are also igneous intrusions: sheet intrusions concordant with the bedding (ie sills) have outcrops parallel to those of the other bedded rocks, whilst those which are discordant (dykes) cut across them and are often vertical; large intrusions of coarse-grained igneous rocks (eg granite, gabbro) will form massive outcrops, and the metamorphic aureole will be wide enough to represent on the map (Figure 11.6).

(*h*) **Metamorphic rocks** occur in the aureoles around large igneous intrusions, or cover large areas where the rocks were once deeply buried and regionally metamorphosed but have since been exposed by erosion. No unaltered sediments or igneous rocks can be older than the gneisses or schists of such areas.

(*i*) **'Drift' deposits** include river gravels and alluvium, raised beach materials and glacial deposits: they are normally thin, unconsolidated and of very recent formation compared with the 'solid' rocks. Chapter 5 provides ideas as to the ways in which the history of such deposits can be worked out.

Geological History. Having studied the outcrop patterns, a general geological history of the area can be deduced, working from the oldest to the most recent events. There will often be half a dozen or more events to sort out.

(*a*) If the order of the rocks outcropping in the area is not given, work it out using the dip arrows and unconformities and, if necessary, drawing strike lines.

(*b*) Then proceed to work out the order of intrusions, folding and faulting. Each will only affect the rocks formed before them, and can be dated as younger than the rocks they affect, and older than any masking unconformity. For instance, a fault or intrusion that cuts across and displaces another is more recent in age.

(*c*) Finally the drift deposits should be put in order if this is possible: they will often refer to glacial events or a change in sea-level.

Try this on the maps printed earlier in this book (Figures 8.23, 11.6, 12.9, 12.10, 12.15, 16.3, 17.9).

Techniques. Certain techniques are also useful in analysing maps.

(*a*) Working out the **directions of strike and dip** is fundamental to examining the structures affecting the rocks. This is possible where maps have relief contours and outcrop patterns. Strike lines are drawn, as in Figure 12.4, and dip directions are at right angles to these. It is important to remember that strike lines must pass through bedding plane/identical contour height intersections: two such intersections are necessary in order to draw the first strike line for any rock surface, but as strike lines are parallel on a rock surface dipping at a consistent angle, later lines need to be drawn through only one point.

(*b*) **Completing outcrops.** This type of exercise is related to the original construction of geological maps from field evidence. A few outcrops and their details are given, plotted on a relief map. Strike lines must be drawn through bedding planes at the same height, and a pattern extended over the whole map area. Then bedding plane lines can be drawn through the strike line/relief contour intersection (it often helps to put a circle at each of these intersections). When the bedding plane outcrops have been plotted, the rock-type symbols can be used for shading in the areas between.

(*c*) **Three-point problems** involve a recognition of the three-dimensional nature of geological situations. Once again a relief map is provided with borehole data from three (or more) points. The depths given must be related to OD (or sea-level), and strike lines drawn so that the map can be completed or a solution given.

(*d*) **Drawing geological sections.** A section often serves to summarise the structures and succession of rocks in an area. The most useful sections will therefore be along the line of dip so that they will show the structures most characteristically.

(i) Having chosen the line of your section, you must draw the relief section first.

(ii) Mark the outcrops of rock where they come to the surface on the relief section.

(iii) Plot the position of vertical rocks and structures, such as batholiths and dykes. Although it rarely occurs in practice, you will find that many faults marked on simple geological maps are drawn as straight lines, and are therefore vertical.

(iv) The most accurate way of completing the section is to draw the strike lines of certain bedding planes until they cut the line of section on the map. Mark at least two for each bedding plane, and then they can be plotted on the section: do not worry if the points come above the level of the land, as you need only draw the bed of rock up to the land surface. Similarly, do not be afraid of continuing the section below sea-level—the rocks do not stop there. Put in the beds above the uppermost conformity first, and the unconformity itself; then work downwards.

(v) If there are no contours on the map you will probably be told the direction and amount of dip, and you can plot this information on the section using a protractor with the straight edge at the top (ie measuring the dip from the horizontal downwards).

(vi) The finishing touches include correct labelling, marking the vertical and horizontal scale, and locating the section.

Exercises related to the outcrop patterns and the use of these techniques are found in *Geological Map Exercises* by M J Bradshaw and E A Jarman.

Further Reading List

A number of books on geology can be obtained in relatively cheap paperback editions:

M J Bradshaw and E A Jarman. *Reading Geological Maps* and *Geological Map Exercises*, The English Universities Press Ltd, 1969.

F H T Rhodes, H S Zim, P R S Shaffer. *Fossils*, Paul Hamlyn, 1965.

H S Zim, P R Shaffer. *Rocks and Minerals*, Paul Hamlyn, 1965.

F H T Rhodes. *The Evolution of Life*, Penguin Books, 1962.

G Dury. *The Face of the Earth*, Penguin Books, 1959.

W G Fearnsides and O M B Bulman. *Geology in the Service of Man*, Penguin Books, 1961.

A E Trueman. *Geology and Scenery in England and Wales*, Penguin Books, 1949.

L D Stamp. *Britain's Structure and Scenery*, Fontana Books, 1960.

J F Kirkaldy. *The Study of Fossils*, Hutchinson Educational Press, 1965.

D V Ager. *Introducing Geology*, Faber and Faber, 1962.
British Palaeozoic Fossils, British Mesozoic Fossils and British Caenozoic Fossils, British Museum (Natural History), 1964, 1962 and 1959.

J F Kirkaldy. *Geological Time*, Oliver and Boyd, 1971.

F W Lane. *The Elements Rage*, Sphere Books, 1968.

F W Dunning. *Geophysical Exploration*, HMSO, 1969.

The following are Hamlyn all-colour paperbacks:

B Cox. *Prehistoric Animals*, 1969.
I O S Evans. *The Earth*, 1970.
M H Day. *Fossil Man*, 1969.

The following series, published by Prentice-Hall, provides an excellent geological library, and each volume is available in paperback.

W A Ernst. *Earth Materials*, 1969.
A L Bloom. *The Surface of the Earth*, 1969.
B J Skinner. *Earth Resources*, 1969.
D J Eicher. *Geologic Time*, 1968.
L F Laporte. *Ancient Environments*, 1968.
A L McAlester. *The History of Life*, 1968.
K K Turekian. *Oceans*, 1968.

The following titles take the reader farther into the subject, the last four referring him to the most advanced works, and keeping him abreast of developments in the subject. Some of them are available in paperback editions.

J F Kirkaldy. *Minerals and Rocks in Colour*, Blandford Press, 1963.

R Casanova. *Fossil Collecting*, Faber and Faber, 1960.

G W Himus and G S Sweeting. *The Elements of Field Geology*, University Tutorial Press, 1955 (2nd Edition).

H H Read and J Watson. *Introduction to Geology*, Volume 1: Principles, Macmillan, 1962.

A Holmes. *Principles of Physical Geology*, Nelson, 1966 (2nd Edition).

G M Bennison and A E Wright. *The Geological History of the British Isles*, Arnold, 1969.

I G Gass, P J Smith and R C Wilson (Eds). *Understanding the Earth*, Open University, 1971.

Index

This index can be used in three main ways:

1 To refer to the definition or explanation of technical terms used in geology.
2 To find examples of the occurrence of rock types, landforms, etc.
3 To look up the references to a certain topic, the details of which are scattered throughout the book.